Geometric Design

An Artful Portfolio of Mathematical Graphics

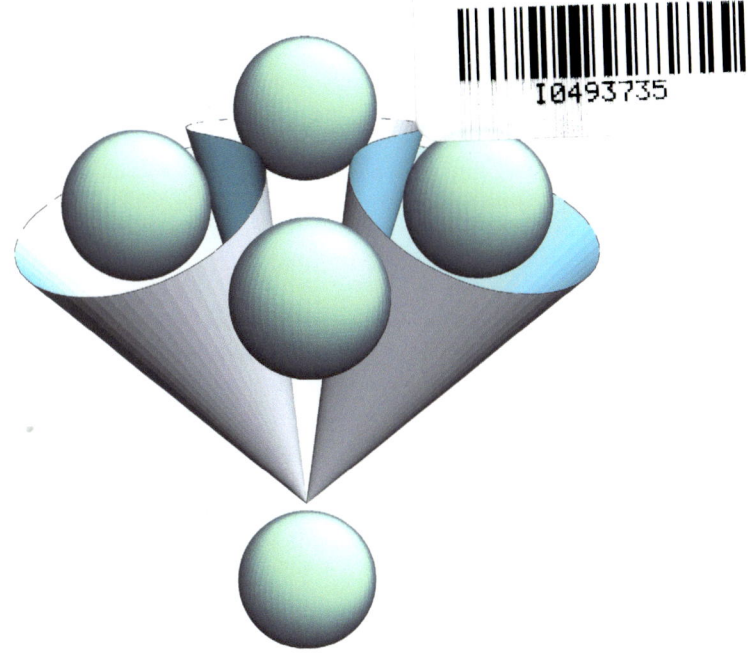

By Christopher Alan Arthur

Requests for copy permission should be mailed to the following address: Permissions, PO Box 1434, Allen, Texas, 75013.

First Edition

Library of Congress Control Number: 2014914096
C. A. Arthur Publishing, Allen, Texas
QA 501.5 .A73 2017

Mathematics Subject Classification (2010): 51-04, 53A05

Printed by CreateSpace, Charleston, SC

Available from Amazon.com and other retail outlets.

Typeset in Minion Pro with *Mathematica 5.2*

Mathematica is a registered trademark of Wolfram Research, Inc.

ISBN 978-0-692-26234-4

First Printing, August 2014

http://www.hythlos.org

Geometric Design Contents

	Introduction	v
1	Torsion I	1
2	The Twisted Square	3
3	The Recursive Pentagram	5
4	Torsion II	7
5	Toroid	10
6	Dodecahedral I	13
7	Dodecahedral II	16
8	One – Pointed Folding	19
9	Torsion III	22
10	Conformal I	24
11	Conformal II	27
12	Aperiodic I	29
13	Aperiodic II	31
14	Concavity I	34
15	Polar Plots	36
16	Triacontahedron	39
17	Cuboid	41
18	Modules	45
19	Squiggle	47
20	Concavity II	49
21	Concavity III	52
22	Self –Similarity I	55
23	Labyrinthine	58
24	Self – Similarity II	60
	References	67
	Glossary of Symbols	68
	About the Author	72

Introduction

Aristotle once argued that a mathematician differs from a physicist, even though they both would discuss the shape of things in the natural world. He claimed that the mathematician would separate the qualities of physical bodies from the things themselves, whereas the physicist (presumably) would not. What I have done in this book is even more remote from the physical world than what Aristotle supposed for mathematicians. I discuss the shape of things not derived from anything at all to do with the real world but rather of whatever has happened whimsically to please my imagination. In *Physics* he goes on to say about geometry that it "investigates physical lines but not qua physical." I believe that to produce this book I have also investigated in this way because the illustrated figures do seem at least potentially to have physical form in the sense of attributes like length and width. Hence, in a philosophical sense I suppose that it is a book of geometry more than anything else.

With regards to what kind of book this is, I wish to point out that even though there are mathematical equations and geometric constructions, it is not a math book like most college texts used for teaching courses. It does not present any proofs of theorems, nor is there much derivation from elementary principles leading to advanced techniques and concepts. It is also not like a doctoral dissertation meant to advance in any singular and profound way the collective pool of human knowledge. I do not intend at all to assert that this book is some bold and serious product of mathematical genius with ubiquitous, scientific implications that any expert must examine. Instead, I feel that the book is in some sense simply my testimony of how mathematics can be something to enjoy, if nothing else. My sincere hope in attempting to publish is that a reader such as yourself might also come to agree with me on this point.

Hence, the goal of the book is to explain and to show some special drawings, but what makes them special is that they are purely mathematical and quite a bit more elaborate than any primitive construction in recollection of high school geometry class. Hopefully, my pictures are also more appealing. Accompanying each drawing there are a few equations and a description of what they mean, plus the stipulation that the reader could use them to produce his or her own version with a degree of modification. This book might therefore be of interest to animators, computer artists or other artists who would want some idea of how mathematics relates to their craft, and it might also be of interest to mathematicians and engineers who could glean from it some new ways of showing their own formulas. Anyone else who might have a general interest in looking at math in a visual way would also find this book worthwhile.

The mathematical concepts applied in this book should be accessible to someone who has studied math at the college level; particularly, a junior who has a major in the physical sciences, engineering or mathematics (of course) should not have much trouble. The course prerequisites might read something like this: college algebra, trigonometry, three semesters of calculus up through multivariable calculus, linear algebra, set theory or discrete math, the theory of a complex variable, and possibly a course in dynamical systems.

The images in this book are obviously not drawn by hand, as the visible level of detail and precision would imply just by our looking at them. They have been produced by a modern computer using software which differs considerably from artistic software generally in that it is a programming language interpreter. The user types in commands and the computer executes them. For example, the six lines of code shown are written in the *Mathematica* language, and they produce a red, elliptical picture. The code is an elementary example of the kind of programming required to produce the images in this book.

```
f[{r_,θ_}]={r Cos[θ],r Sin[θ]};
T=Table[{r,θ},{r,0,1,1/10},{θ,0,2π,π/10}];
A=Map[f,T,{2}];
P=Partition[A,{2,2},{1,1}];
M=Map[Polygon@Drop[Flatten[#,1],1]&,P,{2}];
Show[Graphics@{Red,M}];
```

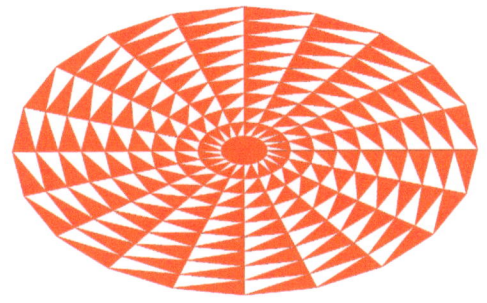

Here is a brief explanation of what the code means. The first line defines a function *f* of two variables *r* and *θ*. The second line makes one set *T* that is the product of the two sets {0, 1, 2, ..., 10} and {0, 1, 2, ..., 20}, such that the ordered pairs in *T* take the values (*a* / 10, π *b* / 10) with *a* and *b* taken from the first and second sets respectively. The next line computes *A* = *f*(*T*) as the image of set *T* through the function *f*. The following line regroups the elements of *A*, which have had the same structure as *T*, and arranges them into sets of four, picking elements in adjacent rows and columns. The fifth line declares that each set of four points should become a triangle by dropping the first point from each set, and the last line tells the computer to display all the triangles together in a shade of red.

The more advanced features of the software that the computer applies to these images include commands for solving linear systems of equations, symbolic and numerical differentiation and integration, matrix and vector calculations, and the projective geometry related to calculating perspective (such as vanishing-point perspective).

Several of the examples take ideas from a book entitled *The Harmony of the World*, written by Johannes Kepler. Although it was originally written in Latin, the American Philosophical Society published a nice English translation by Aiton, Duncan and Field in 1997. In this book Kepler tried to assert that the planets of the solar system move in harmony like music because of their elliptical orbits and oscillating angular velocities. In order to make his assertion he began with some facts about plane geometry and took a musical journey going finally to astronomy. Along the way he also made some appealing observations about polyhedra and plane tessellations, and with them he was arguing that the factors in their construction were abstractly harmonic, as the same ones could be found in musical tuning systems of the day. So, the cosmographical geometry finally is—as he asserted—also somehow harmonic.

In trying to understand Kepler as my thesis in graduate school, I took a liking to some of his illustrations. My fondness is why I have chosen to reconstruct some of his ideas, and to do so was a fair effort in analytical geometry, such as to construct with a computer model a rhombic dodeca-hedron (*Dodecahedral I*) given just a simple sketch and the ratio of diagonals. The construction was a few hours of work.

At points in the text when I make use of ideas from *The Har-mony of the World,* I will refer to it as Kepler's book.

Here are some brief descriptions of the sections in this book.

Torsion I is a patterned, solid figure like a helix wrapped around an invisible torus. The figure is the tessellated boundary of a tubular neighborhood for a simple, closed and differentiable space curve with torsion.

The Twisted Square is a checkered plane figure curled increasingly towards its frontier and is also the image of a dynamical composition of transformations defined by a vector field.

The Recursive Pentagram is a fractal plane figure constructed by subdividing a five-pointed star. The pentagram is a self-similar set of line segments generated from the recursion of affine transformations.

Torsion II shows another solid figure along with a template provided for its patterned surface. Like *Torsion I*, the figure is also the tessellated boundary of a tubular neighborhood for a simple, closed and differentiable space curve with torsion.

Toroid presents mutations of the standard torus, also along with a surface pattern template. The two radii become functions of the angles of the circles themselves, thereby expanding or warping the figure non-uniformly.

Dodecahedral I is a regular, star dodecahedron having a plaid texture, and a series of rotations applied to a pentagonal pyramid can generate the figure.

Dodecahedral II is a regular, rhombic dodecahedron composed of spirals, and it can tessellate a volume of space as a beehive could. The section also gives spatial coordinates for every vertex and a graph of edge connections.

One-Pointed Folding is my attempt to use non-linear paper constructions to interpret mathematically a spiritual concept. The shape is like a cone in having just one sharp singularity, except that this construction is neither circular nor technically even a conic section.

Torsion III shows variations of a spherical, spiral tube also tapered at the ends. Like *Torsion I* and *Torsion II* it is the tessellated boundary of a tubular neighborhood of a space curve that is simple, has torsion and is smooth, but unlike the others, it is not a closed curve.

Conformal I is a pair of related, complex plane figures from rational functions. They are continuous images of a disk and examples of what the *Riemann open mapping theorem* suggests.

Conformal II is a complex planar image approximated by a pattern of crosses and circles. The image exhibits the sort of bipolar curvature typical of electrostatics.

Aperiodic I is a linear, spatial figure related to a "generalized Lissajous" figure except that it is more irrational. I was inspired to make it as a visual image illustrating the difference between musical tuning systems such as equal temperament, being irrational, and just intonation, being rational.

Aperiodic II is set of a plane figures as mostly irregular patterns of colored stripes. The figures are "density plots" of two-variable functions composed with the cosine and relate abstractly to topographical maps.

Concavity I is a collection of line plots showing fivefold and fourfold symmetries. They are smooth, closed and simple curves in the plane.

Polar Plots is also a set of composite line plots with appealing and unexpected turns. The curves are smooth, closed and planar but not all simple.

Triacontahedron is a construction of the thirty-sided, regular and rhombic polyhedron, plus a spline-based figure exploiting its symme-tries. The section includes a table of vertex coordinates and a graph of edge connections.

Cuboid is like a skeleton of cube edges which have been thick-ened into a solid. The section presents a solid figure as an implicit surface and also in approximation a parameterized, similar surface.

Modules is a periodic, spatial figure from a sum of cosines decomposed into hexagonal regions. The discussion first states an implicit equation later solved and integrated to compute volume.

Squiggle shows recursive line drawings generated from smooth curves. They are superimposed upon each other such that the previous curve becomes the frame of the next one.

Concavity II is a set of symmetric solids having varying numbers of cusps and having computable volumes. The solids have differentiable and non-convex parametric surfaces derived in spherical coordinates.

Concavity III is the boundary of a solid with a rather sophisti-cated curvature. A composition of spherical coordinates defines the figure, which is also continuously differentiable.

Self-Similarity I is a solid like a tree with a threefold branching and with questionable dimension. Recursive applications of affine transformations upon a generating set produce this figure.

Labyrinthine is an interesting style of surface illustration applied to familiar shapes. The style is a replacement for a rectangular mesh used for manifold approximation with computation.

Self-Similarity II is an aggregate of cubic patches forming a sort of cruciform tree with a fivefold branching. Computation shows that for the figure extended ad infinitum, the volume converges while the surface area diverges.

1 Torsion I

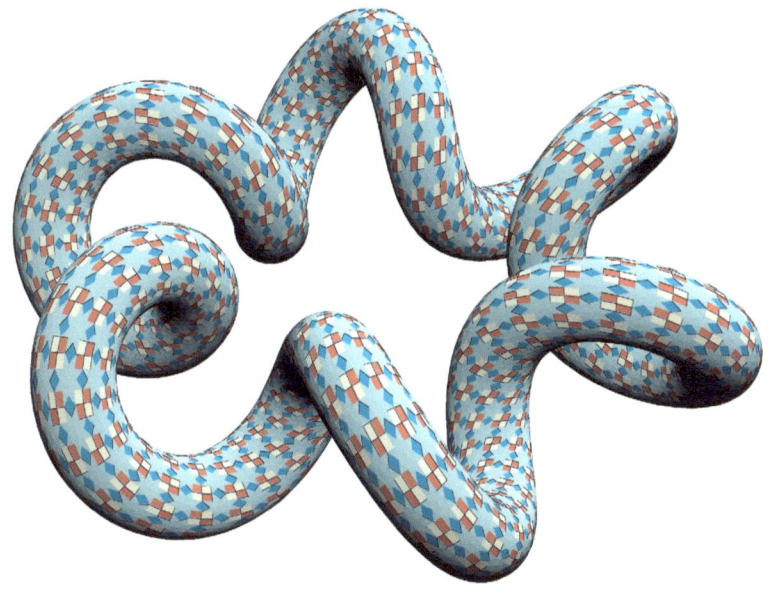

Figure 1-1

Although Figure 1-1 looks like a curly tube, there is still something perfectly circular about it: the cross-section. Just as all circles have center points, then so imaginable is a curvy, continuous line made up solely of such points inside the tube. All of them lie on the surface of a torus with major radius $R = 3$ and minor radius $r = 1$. Here are the general equations for the torus:

$$\begin{aligned}
x(\phi, \theta) &= (R - r\cos\theta)\cos\phi \\
y(\phi, \theta) &= (R - r\cos\theta)\sin\phi \\
z(\phi, \theta) &= r\sin\theta
\end{aligned}$$

The coordinate ϕ signifies a rotation around the major circle, and θ is for the minor one. By letting $\theta = 6\phi$ and $\phi \in [0, 2\pi)$, the torus becomes the line $\sigma(\phi) = (x(\phi), y(\phi), z(\phi))$. It is effectively a loop with no beginning and no end, so it is a *closed* curve. Because it never crosses the

1

same point more than once, it is a *simple* curve. Topologically the loop is in fact homeomorphic to the unit circle, but unlike the circle, it has torsion because it is also a bit like a helix.

A good sense of direction helps to chart the neighborhood of space near and around the curve σ. The lengthwise vector pointing along the curve is $T = \sigma'(\phi)$. An outward direction is given by $n = T'(\phi)$, and another outward direction that is normal to the first is $B = T \times n$. The three together form an orthogonal basis, in terms of which $F(\phi, \theta)$ describes the surface of the tube with ϕ, $\theta \in [0, 2\pi)$ and with both $\hat{n}$ and $\hat{B}$ normalized from n and B.

$$F(\phi, \theta) = \sigma(\phi) + \left(\hat{n}(\phi) \cos \theta + \hat{B}(\phi) \sin \theta\right) / 2$$

Tessellations are commonly patterns that fill up the plane when repeated, and a rectangular part of it can form a torus when rolled up in two ways. Since this figure is homeomorphic to a torus, so can such a pattern readily wrap around it. The particular design here appeared in Kepler's book. Essentially in the design there are six squares for every star octagon, and within each star there are sixteen isosceles triangles similar to $\triangle ABC$ shown in Figure 1-2. Specifically, $\angle CAB = 45°$, $|AC| = 1$, $|AB| = 1$, and $|CB| = \sqrt{2 - \sqrt{2}}$.

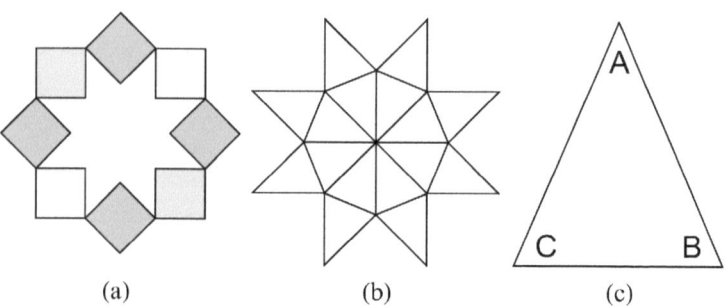

(a)   (b)   (c)

Figure 1-2

2 The Twisted Square

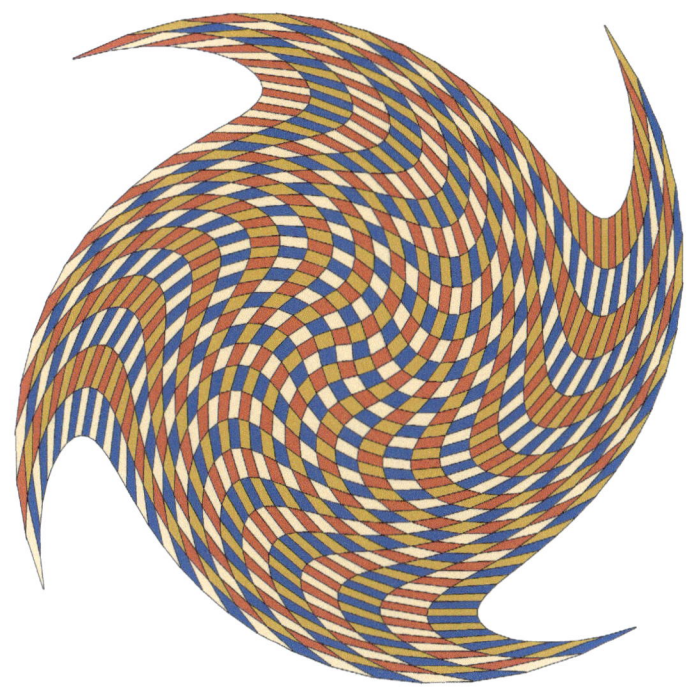

Figure 2-1

The pattern in Figure 2-1 is almost like what happens when a stretchy rubber square with painted stripes turns around in a circle, except that the outside of the square is turning faster than its inside. A vector field can describe such a rotation. Suppose that $p = (x, y)$ is a point on the square with the origin $(x, y) = (0, 0)$ being right in the middle. A small rotation is like the mapping $p \mapsto (-y, x)_p$, meaning that each point turns some amount in a particular direction counterclockwise. The product of that amount times the distance from center

$|p| = \sqrt{x^2 + y^2}$ would turn the outside faster than the inside. A vector field associates a vector v with each point p such that v is also a function of p. In particular $v_p \in (\mathbb{R}^2)_p$ and $(v(p))_p = (-y \cdot |p|, x \cdot |p|)_p$. One limitation to this way of thinking is that it does not describe directly a curving motion but instead breaks it up into several small pushes along straight-line paths. The length of the pushes is reducible by a factor of $m = 1/10$, improving the curvature. Thus a composition F of a twisting transformation $f(p) = p + m\, v(p)$ approximates the apparent rotation of the figure.

$$F_n(p) = \underbrace{(f \circ f \circ \cdots \circ f)}_{n}(p)$$

A series of images shows successive application of function f to the striped square $A = [-2, 2] \times [-2, 2]$ in Figure 2-2. The first image in the upper left corner is a picture of A without transformation, and the last one is for $F_{12}(A)$ with the others in between.

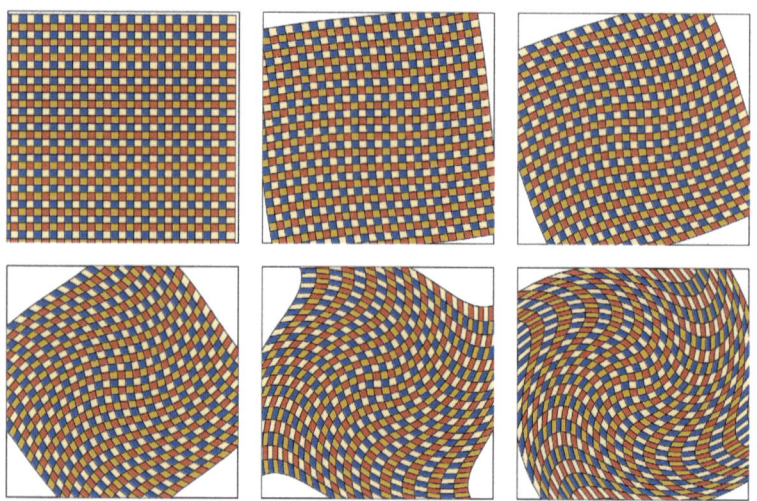

Figure 2-2

4

3 The Recursive Pentagram

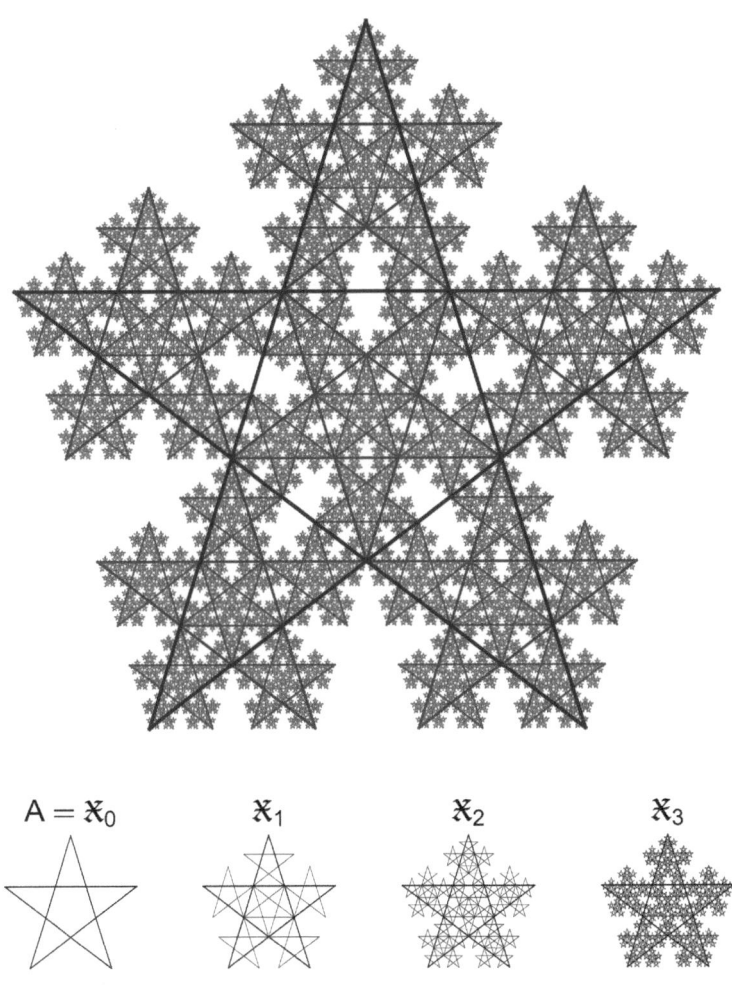

Figure 3-1

In Figure 3-1 there is a self-similar arrangement of five-pointed stars, in which each of the five points of one star becomes a smaller, similar star, which in turn becomes a set of even smaller stars and so on. Actually each one becomes six more because there is another small star placed in the central pentagon. The descendant stars differ from their ancestors by position and size, so that each generation is a set of six affine transformations of the pentagram A. They take the form $Mx + p$, such that

$$M = \begin{pmatrix} u \\ v \end{pmatrix} = \begin{pmatrix} u_1 & u_2 \\ v_1 & v_2 \end{pmatrix} \Rightarrow Mx + p = \begin{pmatrix} u_1 x_1 + u_2 x_2 + p_1 \\ v_1 x_1 + v_2 x_2 + p_2 \end{pmatrix}$$

M is the basis for rotation and scaling, and p represents a translation. Figure 3-2 shows the orientation and location of five bases (the sixth one for the pentagon is not shown).

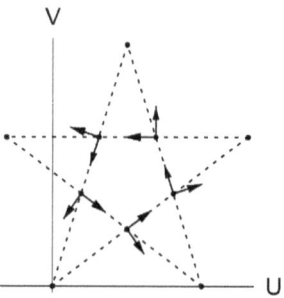

Figure 3-2

Consider the map $A \mapsto \{(M_i, p_i)\}_{i=1}^{6}$ and g_i defined by $g_i(A) = (M_i, p_i)$. Let $\oplus$ denote the application of the affine transformation: $A \oplus (M, p) \equiv \{Mx + p : \forall\, x \in A\}$. In this way the recursion is symbolized by the above sequence with $\hat{i}_n \in Y^n = \{1, 2, ..., 6\}^n$, and the figure $\mathbf{X}_m$ is the union of the recursions up to the level $m = 4$.

$$\mathcal{F}(i) = A \oplus g_i(A)$$
$$(\mathcal{F} \circ \mathcal{F})(i_1, i_2) = A \oplus g_{i_2}(\mathcal{F}(i_1))$$
$$\vdots$$
$$\mathcal{F}_n(\hat{i}_n) = \underbrace{(\mathcal{F} \circ \mathcal{F} \circ \cdots \circ \mathcal{F})}_{n}(\hat{i}_n) \Longrightarrow$$
$$\mathbf{X}_m = \bigcup_{n=0}^{m} \left(\bigcup_{\hat{i} \in Y^n} \mathcal{F}_n(\hat{i}) \right)$$

4 Torsion II

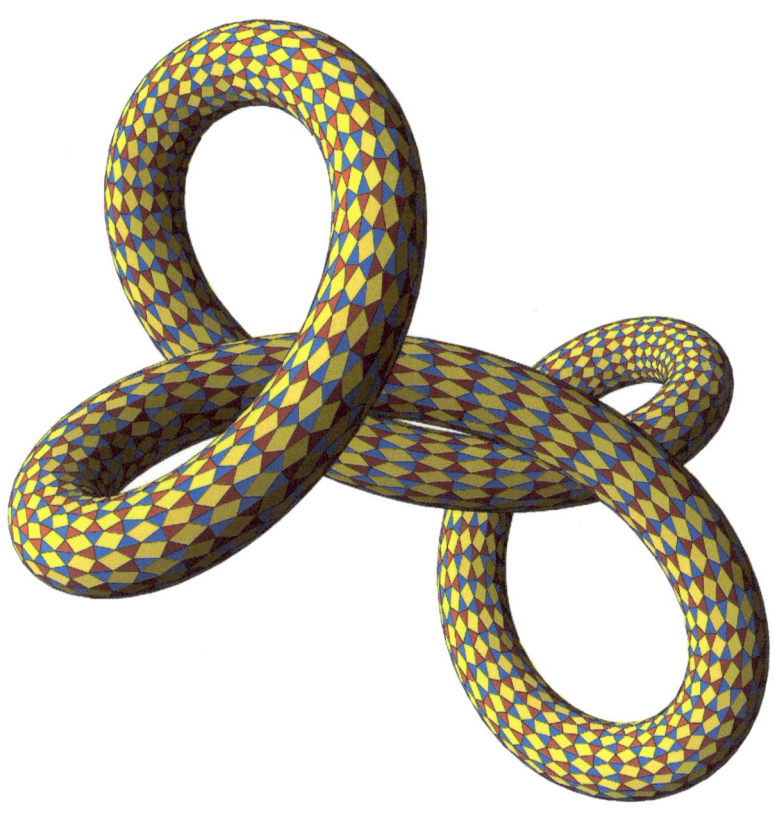

Figure 4-1

Figure 4-1 is similar to the one shown in *Torsion I*, Figure 1-1, since it is the tessellated boundary of a tubular neighborhood for a space curve, and also the curve is simple, is closed, and has torsion. The differences between the two figures are apparent in two ways, by the symmetries of the curve and by the surface pattern. The figure from

Torsion I has six identical loops which are radially symmetric about a circular region in the horizontal plane, whereas the current figure seems to repeat a loop in just two places. Here is the equation for the curve of the current figure:

$$r(t) = 9 \begin{pmatrix} \cos(t) \sin(2\,t) \\ \sin(t) \sin(2\,t) \\ \sin(t + \frac{\pi}{12}) \end{pmatrix}.$$

As before, three orthogonal vectors serve to chart the space near and around the curve. The first vector is $T = r'(t)$ and it points along the path. The other two are $n = T'(t)$ and $B = T \times n$, both pointing outward from the path. The function $F(t, \theta)$ is the entire surface of the tube given parametrically with $\hat{B}$ and $\hat{n}$ normalized and $t, \theta \in [0, 2\pi)$.

$$F(t, \theta) = r(t) + \hat{n}(t) \cos \theta + \hat{B}(t) \sin \theta$$

Since the tube is homeomorphic to a torus, it is also in this way like a flat plane region that is rectangular (and wraps over to meet itself). Hence, it very neatly takes a tessellation, which in this case also appeared in Kepler's book. Two rules generate the pattern from equilateral triangles and squares. First, every square meets a triangle on every edge, and second, every triangle meets two squares plus another triangle. Figure 4-2 gives the coordinates of a set that would tile the plane if repeated, using as constants $\alpha = \sqrt{3}/2$ and $\beta = 1/2$. Following slightly different rules could make different tessellations from the same two shapes. For example, the second rule could instead require that each triangle meets one square and two other triangles instead of just one.

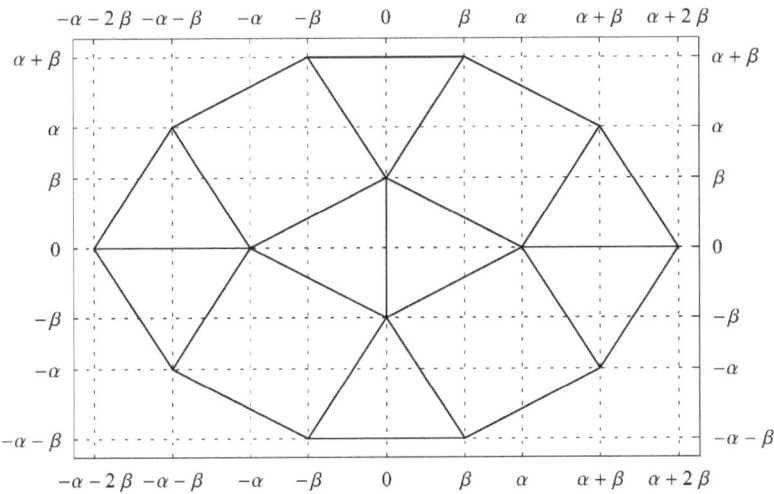

Figure 4-2

Tubular plots such as these could be more interesting potentially if a function would modify the circular cross-section. Restating the formula for *F* in a matrix form might be instructive. A matrix product shows for each of *n* and **B** the three coordinate functions multiplied by a 2×1 matrix with *x* and *y*. If a simple circular section is desirable, then $x(\theta) = \cos \theta$ and $y(\theta) = \sin \theta$, the two equations for a circle, but for instance, they could conveniently become an elliptical figure with something like $x(\theta) = 3 \cos \theta$ and $y(\theta) = 2 \sin \theta$. Later in this book, in the section entitled *Concavity I* there are formulas for other curves which could make interesting cross sections as well.

$$F(t, \theta) = \begin{pmatrix} n_1 & B_1 \\ n_2 & B_2 \\ n_3 & B_3 \end{pmatrix} \cdot \begin{pmatrix} x(\theta) \\ y(\theta) \end{pmatrix} + r(t)$$

9

5 Toroid

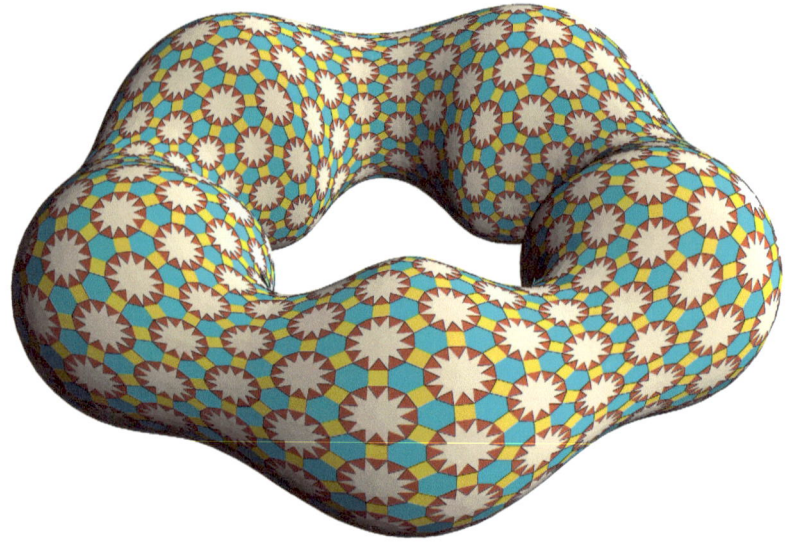

Figure 5-1

Figure 5-1 might look a bit like a toy used for floating in a swimming pool or like an over-inflated inner tube, but geometrically speaking it is essentially a toroid. Notice how some stars are bigger than others because of how the pattern stretches when the shape bulges out. The equations for a torus are

$$\begin{aligned} x(\phi, \theta) &= (R - r\cos\theta)\cos\phi \\ y(\phi, \theta) &= (R - r\cos\theta)\sin\phi \\ z(\phi, \theta) &= r\sin\theta \end{aligned}$$

with $\phi, \theta \in [0, 2\pi)$ and where in this case the major radius is $R = 5$, and the smaller radius is a function of the angle through the larger circle. In particular the function is $r(\phi) = \sin(5\phi)/2 + 2$, which is sinusoidal, and so it moves gradually and repeatedly between fixed upper and lower limits, $r = 5/2$ and $r = 3/2$ respectively. Incidentally ϕ is the major angle, moving through the larger circle, and θ is the minor angle, moving through the smaller circle.

Overall the toroid in Figure 5-1 seems mutable in a couple of ways. For example, the radius functions could change with either of them depending on either or both angles. Instead of $r(\phi)$ as given there could be something like $r(\phi, \theta) = 2 + \sin(5\,\phi)\sin(2\,\theta)/2$, which would have a strange effect (Figure 5-2), but the important point is to maintain certain boundary conditions, such as $r(0, 0) = r(0, 2\,\pi)$, so that the changes create no gaps and that the figure keeps looking solid. Perhaps the major radius could also depend on the angles.

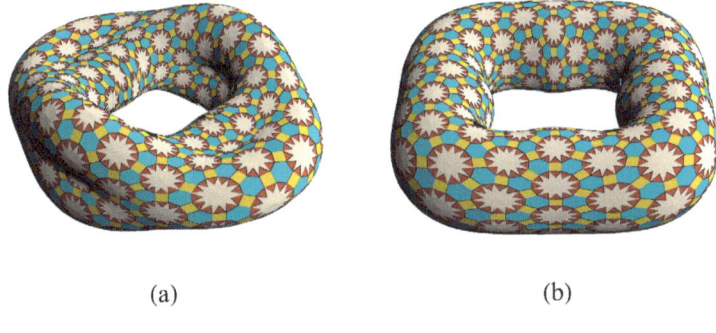

| (a) | (b) |

Figure 5-2

Four regular polygon types form the tessellation pattern, and they are hexagons, squares, triangles, and also a star dodecagon. Like other patterns used in these pictures, this one seems first to appear in Kepler's book. Figure 5-3 shows a diagram plotting the vertices for the shape arrangement.

$$r_1 = \left(\sqrt{2} - \sqrt{3} + \sqrt{6}\right)/2 \qquad a_n = (r_1\, x_n,\, r_1\, y_n)$$

$$r_2 = \sqrt{2 + \sqrt{3}} \qquad\qquad b_n = (r_2 \cos k_n,\, r_2 \sin k_n)$$

$$r_3 = 1 + 3\left(\sqrt{3}\right)/2 \qquad c_n = b_n + a_n/r_1$$

$$x_n = \cos(\pi(n-1)/6) \qquad d_n = b_{n-1} + a_n/r_1$$

$$y_n = \sin(\pi(n-1)/6) \qquad q_n = (r_3\, x_n - y_n/2,\, r_3\, y_n + x_n/2)$$

$$k_n = \pi(2\,n - 1)/12 \qquad p_n = (r_3\, x_n + y_n/2,\, r_3\, y_n - x_n/2)$$

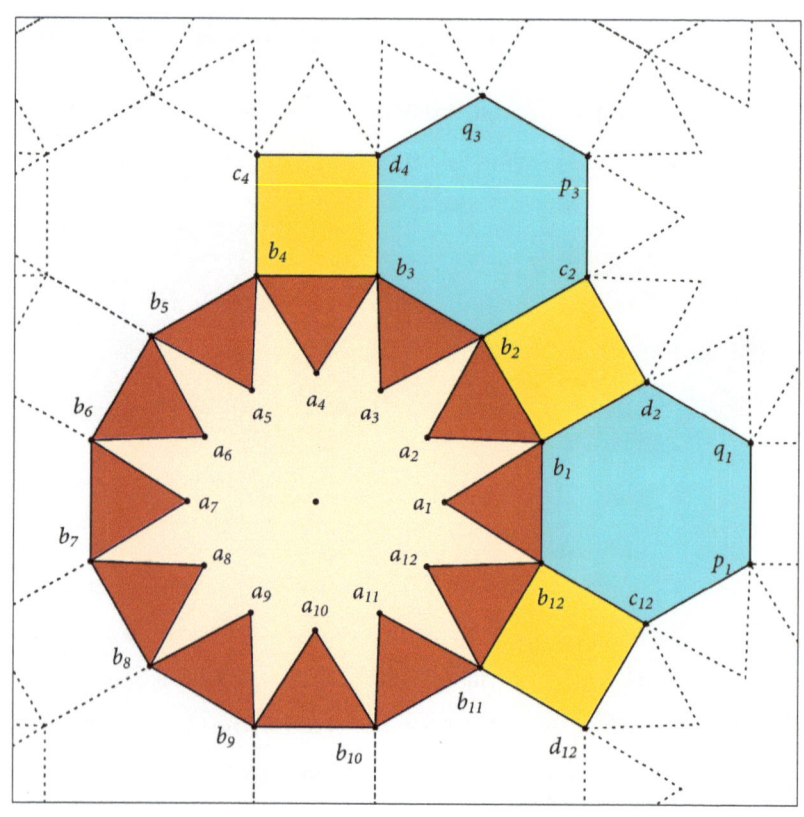

Figure 5-3

6 Dodecahedral I

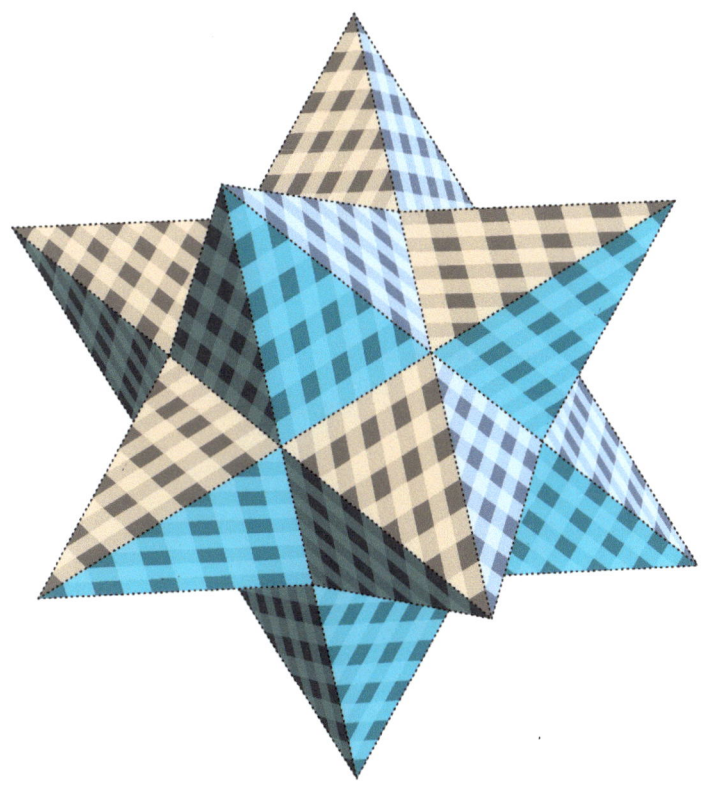

Figure 6-1

The star dodecahedron is a solid that has a set of sixty faces occurring in twelve different planes, each plane containing a five-face subset, which together forms a pentagram. Due to the high degree of symmetry, the whole of Figure 6-1 is constructible from rotations and reflections of a single pyramid in space. In particular, let A be the pyramid having the vertices v_i given by the table. The arrangement of edges between them is shown in the graph with Figure 6-2(a).

$$a = \tfrac{1}{2}\sqrt{\tfrac{1}{2}\left(5 - \sqrt{5}\right)} \quad e = -\tfrac{1}{2}\sqrt{\tfrac{1}{2}\left(5 - \sqrt{5}\right)}$$

$$b = \tfrac{1}{4}\left(-1 - \sqrt{5}\right) \quad f = \tfrac{1}{4}\left(3 + \sqrt{5}\right)$$

$$c = \tfrac{1}{4}\left(-1 + \sqrt{5}\right) \quad g = \tfrac{1}{4}\left(5 + 3\sqrt{5}\right)$$

$$d = \tfrac{1}{2}\sqrt{\tfrac{1}{2}\left(5 + \sqrt{5}\right)} \quad h = \tfrac{1}{2}\left(1 + \sqrt{5}\right)$$

V	X	Y	Z
1	0	0	g
2	1	0	f
3	c	d	f
4	b	a	f
5	b	e	f
6	c	$-d$	f

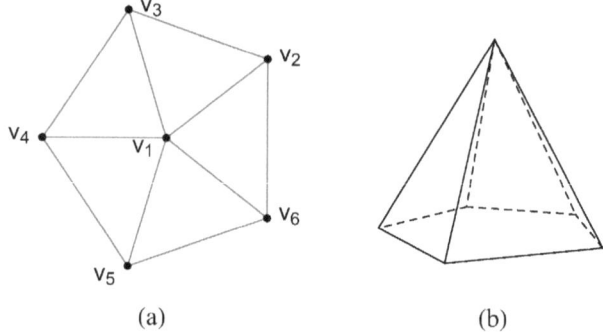

(a)   (b)

Figure 6-2

The upper half of the figure is equivalently the union of A with its image from the linear transformations U and T_k; that is, $B = A \cup \left(\bigcup_{k=0}^{4} T_k \cdot U \cdot A\right)$. Flipping B and joining it back to itself generate the total figure C. Specifically, let $S = (-e_1, -e_2, -e_3)$, where the e_i's are just the usual basis of $\mathbb{R}^3$, and so $C = B \cup (S \cdot B)$. For $c_k = \cos(\tfrac{2k\pi}{5})$ and $s_k = \sin(\tfrac{2k\pi}{5})$ then let

$$T_k = \begin{pmatrix} c_k & s_k & 0 \\ -s_k & c_k & 0 \\ 0 & 0 & 1 \end{pmatrix} \quad \text{and} \quad U = \sqrt{5}\begin{pmatrix} a\sqrt{5} & h/2 & h \\ b\sqrt{5} & a & 2a \\ 0 & -2 & 1 \end{pmatrix}.$$

In order to map some kind of pattern onto the faces, there should be a coordinate system, for which barycentric, homogeneous coordinates would represent the pattern concisely in this case. For three vertices v_{i_1}, v_{i_2}, v_{i_3} making a face, the set of points in the triangle is $\Delta = \{\sum_{j=1}^{3} t_j v_{i_j} : \sum t_j = 1 \, , \, t_j \geq 0 \, \forall \, j\}$. The pattern is then just a shading function $f : \Delta \longrightarrow [0, 1]$, such that zero and one map to two different colors blended in between the bounds of the interval. Also, a pattern of lines parallel to one edge as in Figure 6-3(a) might for example keep t_3 fixed, such that $t_1 = C - t_2$ for some constant C and for $t_2 \in (0, C)$.

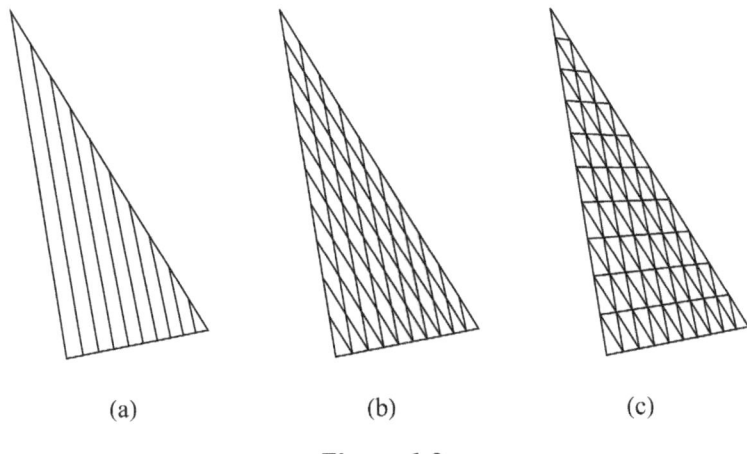

(a)   (b)   (c)

Figure 6-3

Incidentally, I first got a good look at the star dodecahedron in Kepler's book, but probably the shape is much older than that. He had sketched out a small picture of it when he was trying to understand the solid angles that the regular polygons can make. Although I have for convenience presented the locations of the vertices in the table, I had derived them first by putting regular pentagons together, producing the regular dodecahedron, which becomes the star by extension of all the edges until they intersect.

7 Dodecahedral II

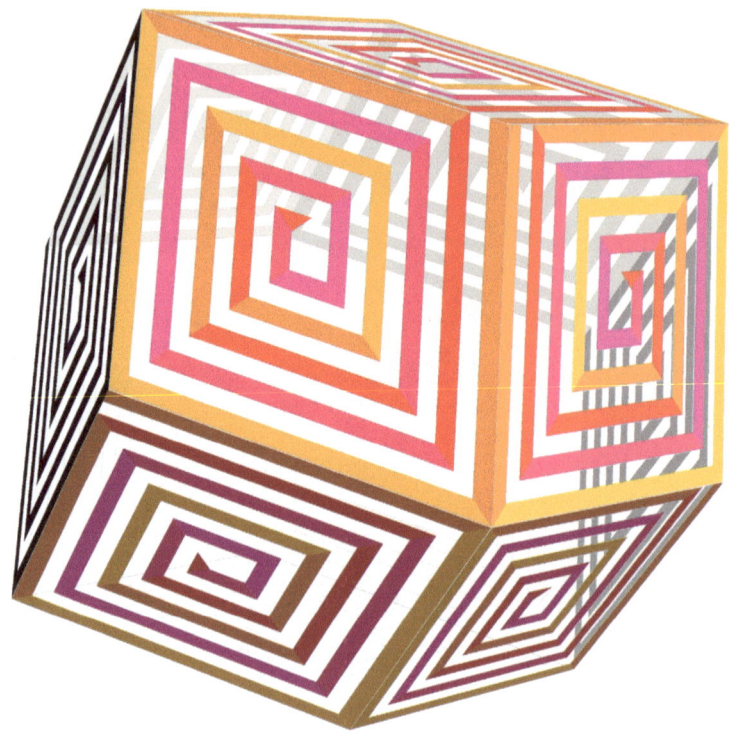

Figure 7-1

Figure 7-1 is an embellishment of a rhombic polyhedron that appears in Kepler's book. The table lists the coordinates for the fourteen vertices, and a line graph in Figure 7-2(a) shows the edges between them. Six faces form a hexagonal loop around the middle, and three faces close off both the top and the bottom. Like a cube, this figure can fit together with itself so to fill space without leaving any gaps. The aggregate with many of this shape has the appearance of a honeycomb.

$$a = \sqrt{\tfrac{2}{3}} \quad b = \tfrac{1}{\sqrt{2}} \quad c = \tfrac{1}{\sqrt{6}} \quad d = -\tfrac{1}{\sqrt{3}}$$

V	X	Y	Z	V	X	Y	Z
1	0	0	$3d/2$	8	$-c$	$-b$	$-d/2$
2	0	0	$-3d/2$	9	$-c$	b	d
3	$-a$	0	$d/2$	10	$-c$	b	$-d/2$
4	$-a$	0	$-d$	11	c	$-b$	$d/2$
5	a	0	d	12	c	$-b$	$-d$
6	a	0	$-d/2$	13	c	b	$d/2$
7	$-c$	$-b$	d	14	c	b	$-d$

Figure 7-2(a) is a graph showing the connections of the vertices corresponding to the enumeration given in the table. The rhombus in Figure 7-2(b) has the aspect appropriate to generate the polyhedron, and the ratio of the diagonals is $1 : \sqrt{2}$.

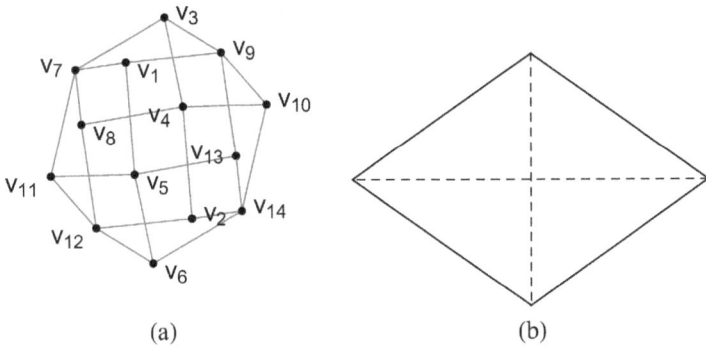

(a)   (b)

Figure 7-2

The path on each face is not properly a spiral since straight segments and not curves are what comprise it. The sequence that builds it is basically starting at the center of the face, moving parallel to an edge, turning left, moving parallel again, turning left, and so on. Each time after a turn the movement is a bit longer and continues until reaching the edge of the face.

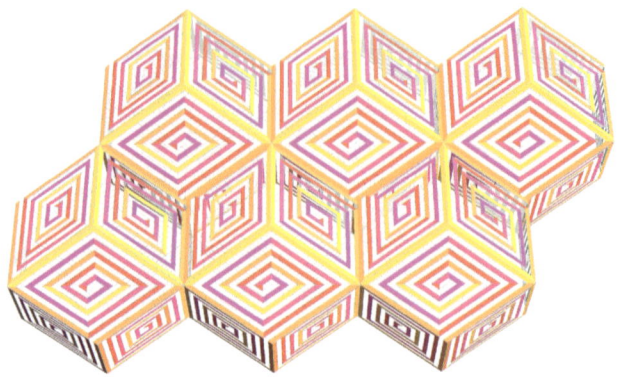

Figure 7-3

Figure 7-3 is a view of how to form an aggregate like a honey-comb using this dodecahedron. With the top three faces removed from each one, the figures would thus have openings bearing resemblance to the cells of a beehive. For the sake of honeyed sweetness, I wonder sometimes if anyone has bothered to make variations on sugar cubes that would take this shape. They could be stacked and packaged in boxes almost as easily as the cubes, but maybe the production machinery manufactures right angles better than it could handle something else like this particularly because cutting several at once—presuming that the cubes are cut—would seem to be more difficult.

Crystallography is, incidentally, the study of shapes such as these insofar as they represent the angles and distances of molecular bonds present in real crystals. In fact the structure of diamond seems to be somehow related to how this shape stacks upon itself.

8 One-Pointed Folding

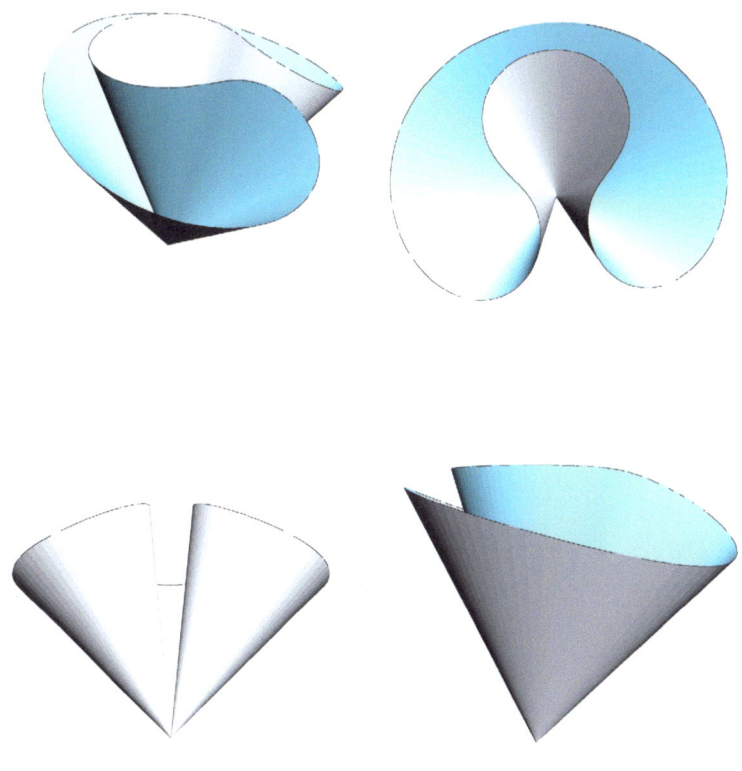

Figure 8-1

The inspiration for Figure 8-1 comes from a mystical idea mentioned in literature by Tibetans, who call it *samadhi*. According to Lama Lodö as he describes in *Bardo Teachings*, samadhi is defined as a quality of "one-pointed involvement in meditation...[and] it does not indicate what the object of meditation is." My interpretation begins with the observation that origami or paper folding can be contemplative and therefore a kind of meditation. The next step is to reduce the lines of folds to just a single point. Hence, I suppose that samadhi is available through contemplative folding of a paper circle into this one-pointed figure.

Here are instructions on how to fold one-pointedly. Set the paper circle down on a flat desk and hold a pencil tip at the center of the circle. Grab the edge of the circle, and while keeping the paper taut, pull it in towards the pencil as the paper starts to bend. Remove the pencil and pinch the sides of the bend inward. The shape should look something like the pictures. Alternatively such a surface is calculable with equations found as follows, starting with these constants:

$$a = \pi + \sin^{-1}(4/5) \qquad m = 2$$
$$b = \pi - \sin^{-1}(4/5) \qquad c = 3/2.$$

Define a spherical coordinate system $F: A \longrightarrow S^2 \subseteq \mathbb{R}^3$ ∋ $A = [0, \pi] \times [0, 2\pi]$ in the following way:

$$F(\phi, \theta) = \begin{pmatrix} \cos\theta \sin\phi \\ \sin\theta \sin\phi \\ \cos\phi \end{pmatrix}$$

To compute the edge curve, let $g(t) = m \, e^{-(c\,t)^2}$, which serves to define a family of functions ϕ_k, each in turn corresponding to the elevation measured from the zenith in a spherical coordinate system, and θ gives the azimuth.

$$\phi_k(t) = k(g(t-a) + g(t-b) + 1)/3$$
$$\theta(t) = \pi - (15\pi/16) \sin(t)$$

Let $H(t, k) = (\phi_k(t), \theta(t))$, essentially interpreted as the space curve homotopy with a parameter k adjusting the shape of the curve. We seek a k_0 that makes the arc length of $H(t, k_0)$ equal to 2π, so that the curve will be equal in length to the boundary of a circular sheet of paper with unit radius. Substituting into the formula for arc length gives:

$$L = \int_0^{2\pi} |\sigma'(t)|\, dt \implies L(k) = \int_0^{2\pi} \left| \frac{\partial}{\partial t} (F \circ H)(t, k) \right| dt$$

Solving the equation $L(k_0) = 2\pi$ numerically gives an approximate value of $k_0 \approx 89017/100000$. To define parametrically the conical shape, let $G : [0, 1] \times [0, 2\pi] \longrightarrow B^3$ such that $G(s, t) = s \cdot (F \circ H)(t, k_0)$ and with $B^3 = \{x \in \mathbb{R}^3 : x_1{}^2 + x_2{}^2 + x_3{}^2 \leq 1\}$. Hence, G gives the figure constructed with the proportions and curvature appropriate to a circular piece of paper.

The picture on the first page of this book embellishes the samadhi-folding with five spheres, and I estimated their size and position from the following description that Lodö gives for an exercise of meditative visualization, as described in his book, *The Quintessence of the Animate and Inanimate.*

The Mahayana practitioner may develop concentration by focusing on an object, for example a small visualized sphere or dot at the level of the nose and ten fingers' distance in front of the face. Concentrating one-pointedly on this, we absorb the mind fully in this object. Once we develop the concentration on this single point, we can add four spheres at a further distance—one in front, one behind, and one each to the left and right. All have the same shape, size, and color. We then focus the mind of these five points at the same time. Once we are able to focus on these five and develop perfect concentration on them, we can then visualize the spheres in the four directions merging into the one in front of us. Then the one in front is seen to grow smaller and smaller until it disappears completely, merging into space.

9 Torsion III

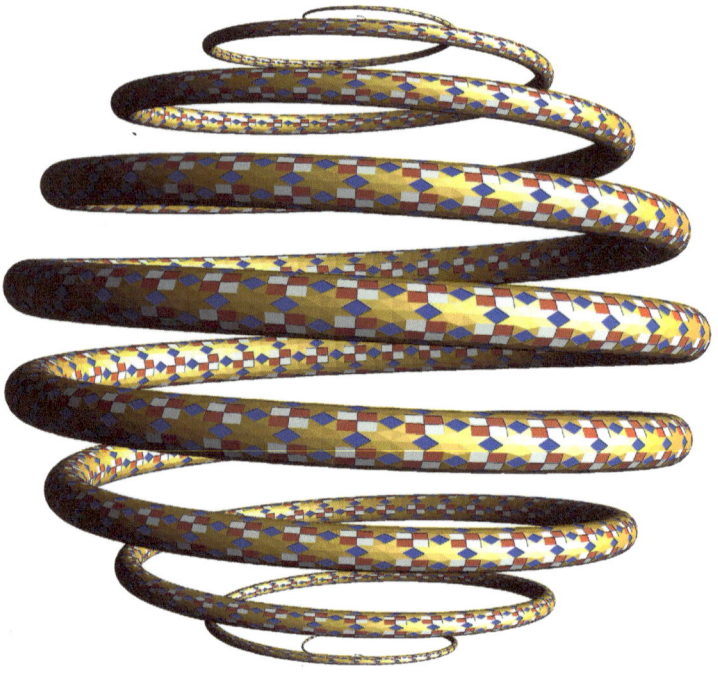

Figure 9-1

The tessellation in Figure 9-1 is identical to the one in *Torsion I*, and this figure also shows a tubular neighborhood of a simple space curve, but the difference is in the shape of the curve, now no longer closed. The equation $r(t)$ for the line through the center of the tube comes from spherical coordinates:

$$f(\phi, \theta) = \begin{pmatrix} x(\phi, \theta) \\ y(\phi, \theta) \\ z(\phi, \theta) \end{pmatrix} = \begin{pmatrix} \cos \theta \sin \phi \\ \sin \theta \sin \phi \\ \cos \phi \end{pmatrix}$$

Define the curve by $r(t) = f(\phi(t), \theta(t))$ such that $\theta = 10\,t$, $\phi = t/2$ and $t \in (0, 2\pi)$. The factor of ten in the assignment for θ determines how many times the curve curls around, and we could increase it for more curls. One consideration with the increase is in how the pattern stretches out as the curve lengthens, and it is likely that the pattern will have to repeat more with length.

As before in *Torsion I*, the three vectors for the neighborhood chart are $T = r'(t)$, pointing along the path, and the others $n = T'(t)$ and $B = T \times n$, both pointing outward from the path. Let $P(t, \theta) = \hat{B}(t) \sin \theta + \hat{n}(t) \cos \theta$ with $\hat{B}$ and $\hat{n}$ normalized, and then let $t, \theta \in [0, 2\pi)$ for $F(t, \theta) = r(t) + (1 - \cos t) P(t, \theta)/k$.

The tube tapers off at the ends, and the tapering obeys the factor $(1 - \cos t)$, which has a kind of bell shape going to zero for $t \to 0$ and $t \to 2\pi$. The mean ratio of tube and sphere radii is the factor $k = 25$. Smaller numbers make a thicker tube, but it overlaps itself if too thick. Replacing $r(t)$ with $r(t) \cdot (2 + \cos n t)/2$ yields curious changes for some $n \in (0, 15)$ as shown in Figure 9-2.

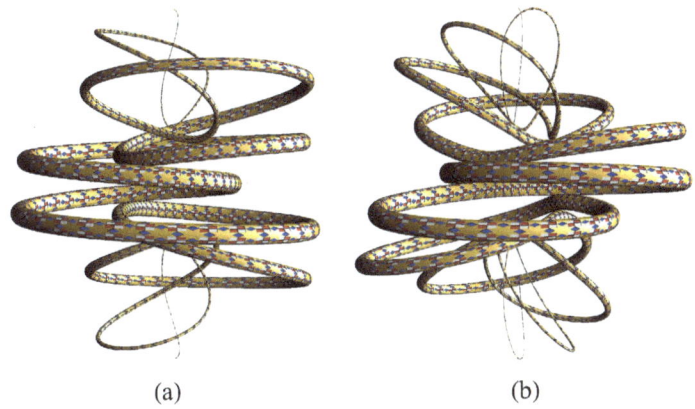

(a)   (b)

Figure 9-2

10 Conformal I

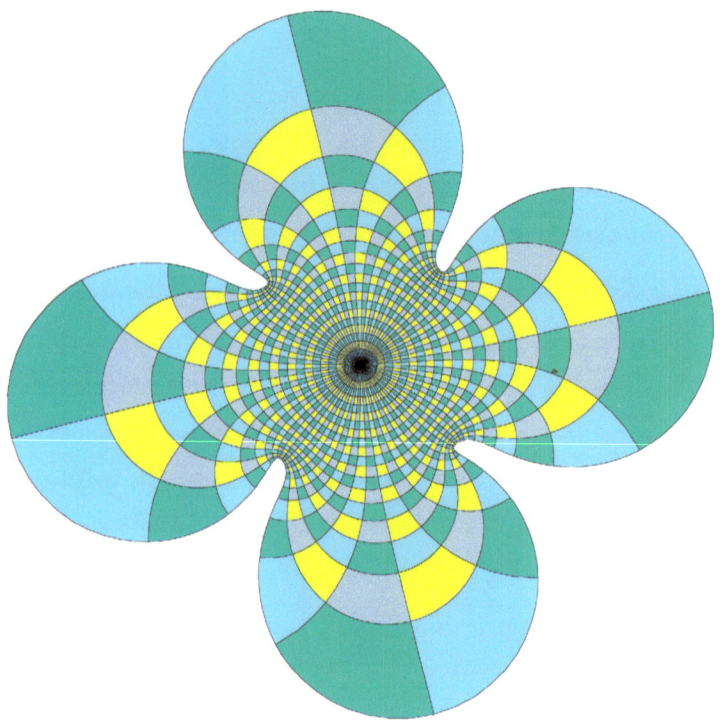

Figure 10-1

The webs of lines in Figures 10-1 and 10-2 illustrate how the functions $f, g : D \subset \mathbb{C} \longrightarrow \mathbb{C}$ are conformal (angle-preserving), and also the figures are the images of the disk D through the two functions. Each is a product of rational expressions as factors which vary by a root of unity. In the case of f there are the fourth roots of unity, and for g, the third roots.

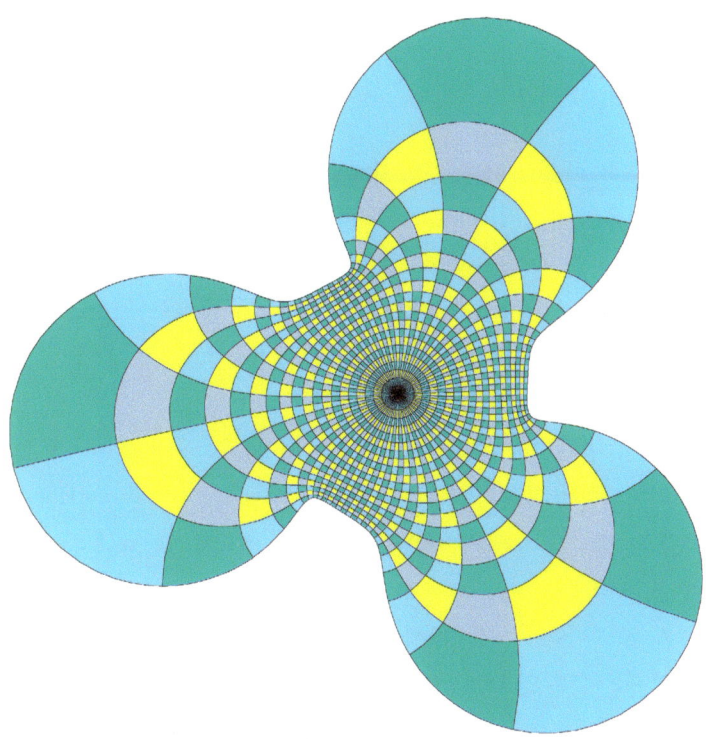

Figure 10-2

$$f(z) = z \cdot \prod_{k=0}^{3} \frac{z + i^k(m+n)}{z + i^k m} \quad ; \quad n = \frac{5}{2} + \frac{i}{2}$$

$$g(z) = z \cdot \prod_{k=0}^{2} \frac{z + u^k(m+n)}{z + u^k m} \quad ; \quad u = \frac{-1}{2} + \frac{i\sqrt{3}}{2}$$

The images of each function are Figure 10-1 with four lobes for *f* and Figure 10-2 with three lobes for *g*. The singularities do not appear because the domains are essentially a disk $D = \{z : |z| < 10\}$ with a radius small enough to miss all the holes; although, it is close to them. In effect, they push and pull on the disk from a distance in order to generate the curvature. Note that if $n = 0$ then f and g are just the identity map, and as long as $|m| > 10$, the two functions are analytic. The images shown take $m = 11$.

Notice that in the center of the figures there are some curves that are almost concentric circles, and they are less and less circular as they move towards the outer edge. This feature relates to the *Riemann open mapping theorem*, which implies that for any simple, smooth and closed curve, there is a conformal map between the interior of the curve and the unit disk. Although, the theorem does not suggest a way to find these conformal maps if just starting with a closed, smooth path. If you are an ambitious reader, then you can try to draw out these maps as an exercise. Start by drawing a smooth, closed and simple curve; i.e., it has to be a loop, but it cannot cross itself or have any sharp kinks in it. Then, draw a small circle somewhere in the interior of the region enclosed by your curve. Draw additional lines extending out from the center of your circle such that they can all bend to meet your curve at a right angle. Next, you can start drawing larger circles concentric with your small one, and the larger they get, the less they should look like circles. Instead, they should look more and more like your curve. At every intersection of lines there should be right angles, too.

With a little practice a modest illustrator can draw the conformal quality reasonably well, provided that the starting curve is not too curvy. When I try this exercise, I find that I get lost and often err when choosing curves other than egg-shaped ones. For any single curve there is in fact only one right answer, but there are a lot of answers that are almost right. Placing the small circle in a good spot makes a difference in how easy the drawing is to make.

11 Conformal II

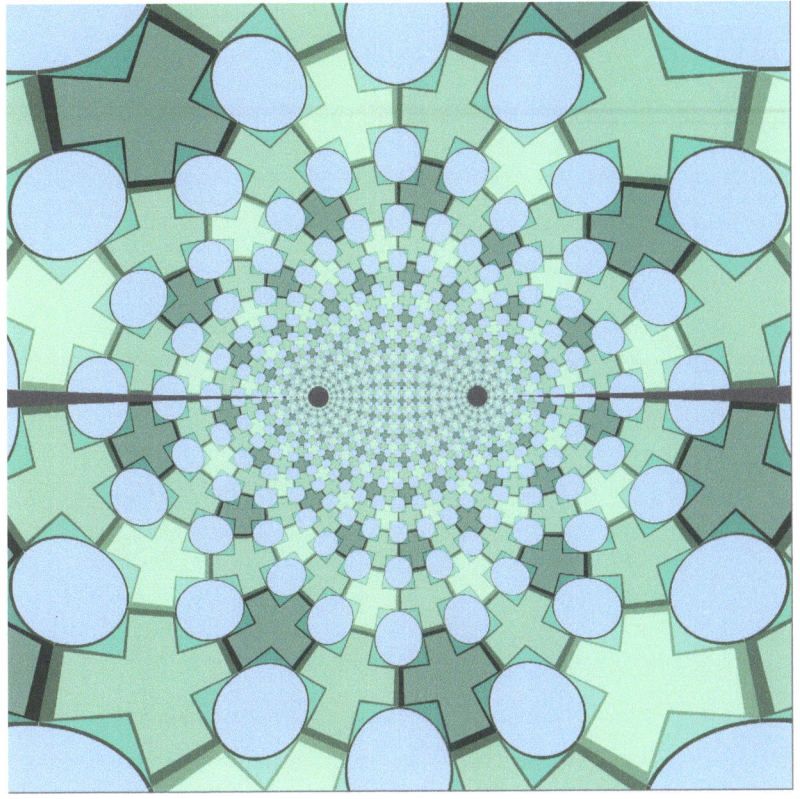

Figure 11-1

A pattern of crosses and circles decorates an image of the hyperbolic tangent through the complex plane in Figure 11-1. Interpreting the shapes and shades as subsets of a box A with side length π, you my reader may note that the pattern repeats as in Figure 11-2, although instead of just three times each way as shown in (b) for simplicity, there are thirty repetitions. Then function f maps A through the tangent, and function g projects the result into the real plane to produce the final picture.

More precisely, let $f : \mathbb{C} \longrightarrow \mathbb{C}$ such that $f(z) = \tanh(z/(10\,\pi))$, and let h be the following linear function serving to shift and stretch the domain $A \in [0, \pi) \times [0, \pi)$ in order to fill in any gaps in the image.

$$h(u, v) = 27 \left(u + \tfrac{29}{25}\, i\, (v - \tfrac{\pi}{2}) - \tfrac{\pi}{2} \right)$$

Finally, let $g : \mathbb{C} \longrightarrow \mathbb{R}^2$ such that $g(u + i\,v) = (u, v)$. Then the picture in the real plane is B in the formula $B = (g \circ f \circ h)\,(A)$.

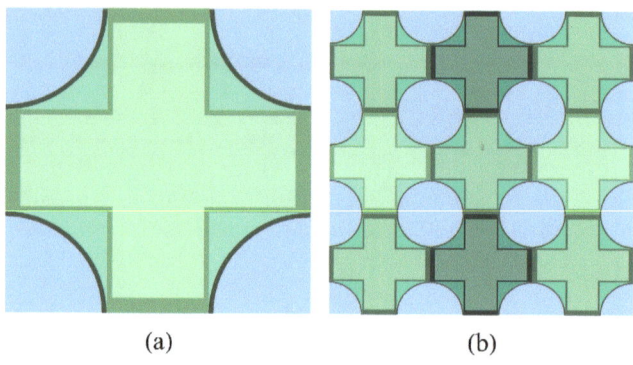

| (a) | (b) |

Figure 11-2

The picture in Figure 11-1 has in fact a slight flaw, in that it is only approximately conformal, since inspection reveals that the right angles of the crosses are not perfectly right in the image. While this fact might not be aesthetically significant, the explanation is that the computer drawing program treated every straight line segment as just a pair of endpoints. In effect, they are in the right spots, but the points between them are not. The program figured that a straight line is the image of a straight line as in Figure 11-2, when it really is not. Nonetheless, where the segments are very short the error is hardly noticeable. One practical alternative would be to tell the computer to consider line segments as a set of many, smaller segments laid end-to-end. The error would still remain, but it would be less apparent.

12 Aperiodic I

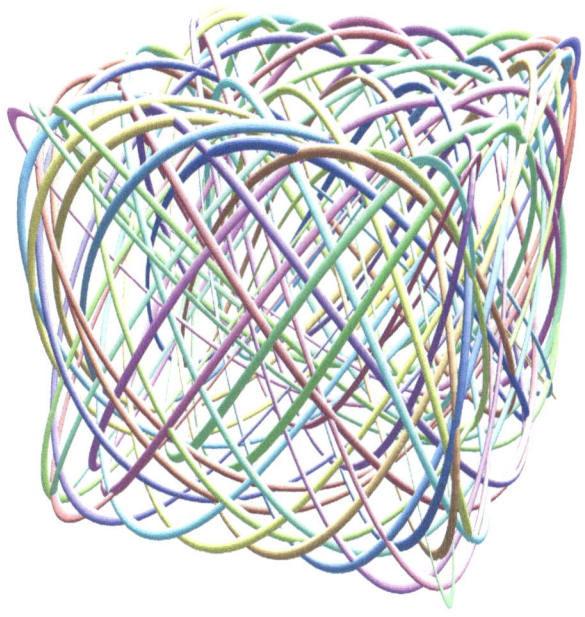

Figure 12-1

Figure 12-1 shows a long length of the space curve formed from a triple of sinusoidal curves. Because each coordinate has a period which is not commensurate with either of the other two, the curve itself has no period and if extended further will continue in a way that seems to fill up the space of the bounding cube. The formula for the curve is generally of the form

$$f(t) = \sum_{i=1}^{3} \sin\left(t \sqrt{p_i}\right) e_i,$$

where $e_1 = (1, 0, 0)$, $e_2 = (0, 1, 0)$, and $e_3 = (0, 0, 1)$. Each p_i is different from the others and is a prime number chosen in this case to be 2, 3 or 5. Although the graph as shown is a sort of tube, the function $f(t)$ is actually just a thin line and could not fill any space. On the other hand, it would seem to be topologically dense in some compact neighborhood of the origin such as the closed unit sphere.

By contrast, closed curves of finite length correspond to commen-surate periods of the coordinate functions. As shown in Figure 12-2 we can for example choose instead of primes for the p_i's, the perfect squares 64, 81 and 144 in (a) and (c) or also 16, 25, and 36 in (b) and (d).

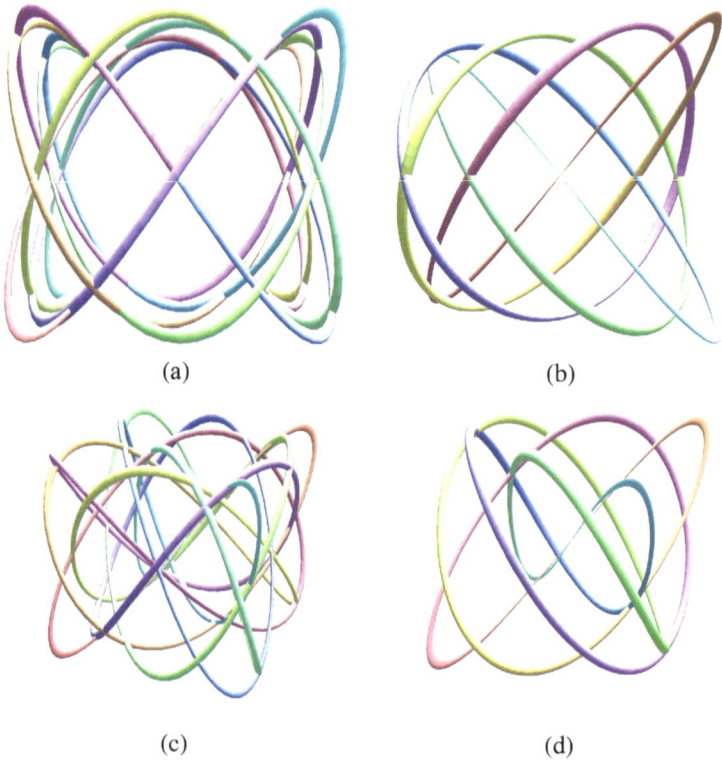

(a)   (b)

(c)   (d)

Figure 12-2

13 Aperiodic II

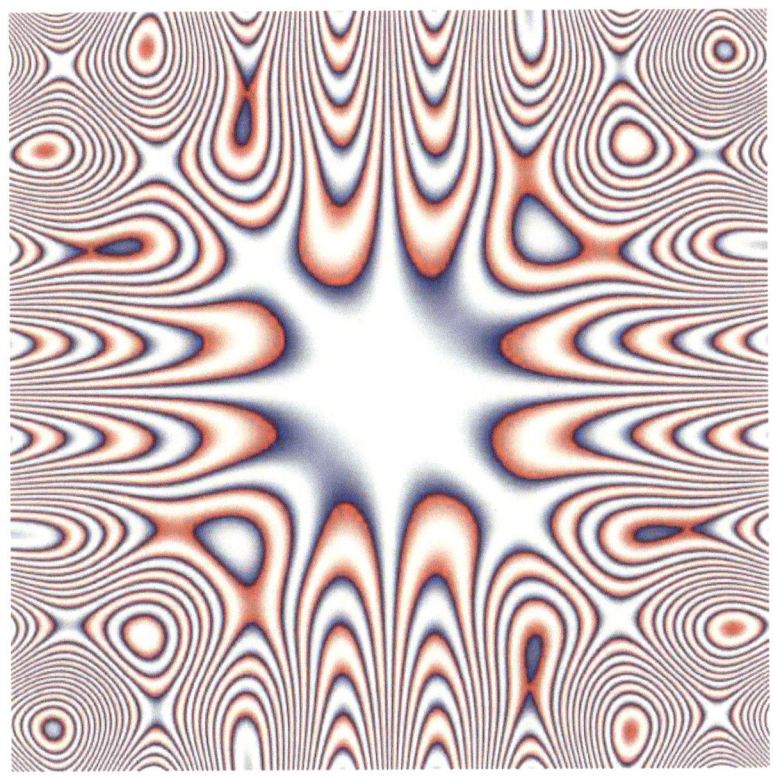

Figure 13-1

Does Figure 13-1 look somehow like a bizarre candy cane or the coat of a dyed zebra? To understand this picture in a sort of technical way would be to think of each stripe like an isopleth or a line of equal heights. For example, where there are many lines close together, there is something like a steep hill on a topographic map.

The effect of contours for a simpler case is shown in Figure 13-2. The figure (a) looks a bit like an egg carton, and it is a three-dimensional plot of $f(x, y)$. The figure (b), being an image of $\cos(\beta\, f(x, y))$, derives from the same functions to the effect of showing the hills and valleys as concentric circles. Incidentally, the circles would be thinner and more numerous if the factor of $\beta = 3\pi/2$ would increase to something larger. How could the same idea be applied to another case with less repetition in the pattern?

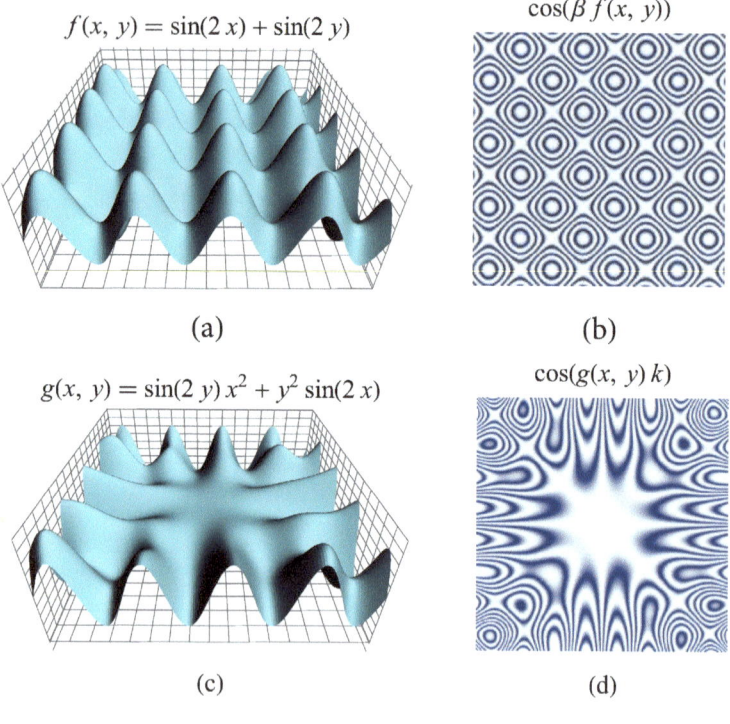

$f(x, y) = \sin(2x) + \sin(2y)$

$\cos(\beta\, f(x, y))$

(a)

(b)

$g(x, y) = \sin(2y)\, x^2 + y^2 \sin(2x)$

$\cos(g(x, y)\, k)$

(c)

(d)

Figure 13-2

As Figures 13-2 (c) and (d) show for $g(x, y)$, this function keeps most of its roots in the same positions as in $f(x, y)$ from (a), but the extrema grow more extreme as they are farther from the origin and from the axes of the graph. As before, the stripes can become thinner and more numerous by an increase in k, which was set to $k = 1/4$ for this picture.

Figure 13-2(d) is less repetitive than (b) but still symmetric across the diagonals. Maybe to improve the design, the goal is to find an asymmetric function without our contriving too much a look of complexity by some meticulous arrangement and sum of bell curves, for example, as shown in Figure 13-3(a).

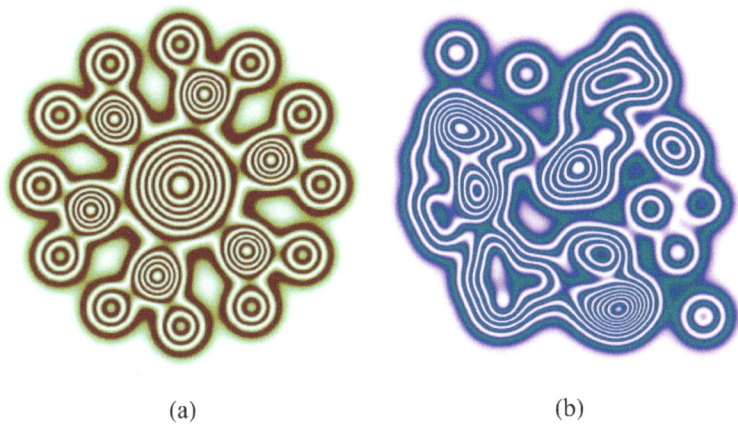

(a)   (b)

Figure 13-3

Both (a) and (b) in Figure 13-3 are similar to the others in the sense that they are a kind of altitude map made from a cosine function of the form $\cos(F(x, y))$, and this time each concentration is an exponential function with a circular power, making a sort of bell curve. Let $C_i(x, y) = (x - x_i)^2 + (y - y_i)^2$, and then $F(x, y) = \pi k \sum_{i=1}^{N} (r_i) 2^{-C_i(x,y)}$. So, the sum ranges over all the different circular centers, each with a specific radius r_i. The difference between (a) and (b), technically speaking, is that (a) has regularly placed circles and (b) has randomly placed ones.

14 Concavity I

These closed curves are all simple and differentiable, and the regions which they enclose are not convex or even star convex, meaning that it is not possible to draw within the region a straight line from the center to all points on the curve because essentially the extremities are too large. The curves are all also polar plots on $f(r, \theta) = (r\cos\theta, \ r\sin\theta)$ such that they are each the graph of a function $C(t) = f(r(t), \theta(t))$ with r as the radius (distance from origin) and θ as the angle (from standard position). Often, polar graphs have r as a function of θ, but in these cases both variables are functions of $t \in [0, 2\pi)$. All of the angle functions are essentially of the form $\theta(t) = t + a\sin(bt)$ where constants a and b are chosen carefully so that θ decreases just a bit in a few places (but is mostly increasing). The sine period is always a multiple of the cosine period given with radius $r(t)$, and both can change with the desired number of branches.

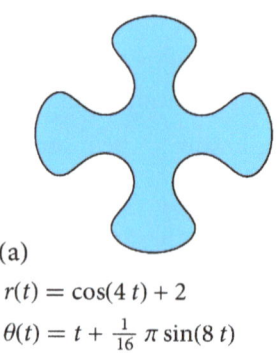

(a)

$r(t) = \cos(4t) + 2$

$\theta(t) = t + \frac{1}{16}\pi\sin(8t)$

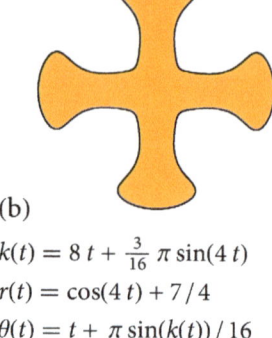

(b)

$k(t) = 8t + \frac{3}{16}\pi\sin(4t)$

$r(t) = \cos(4t) + 7/4$

$\theta(t) = t + \pi\sin(k(t))/16$

 Concavity I

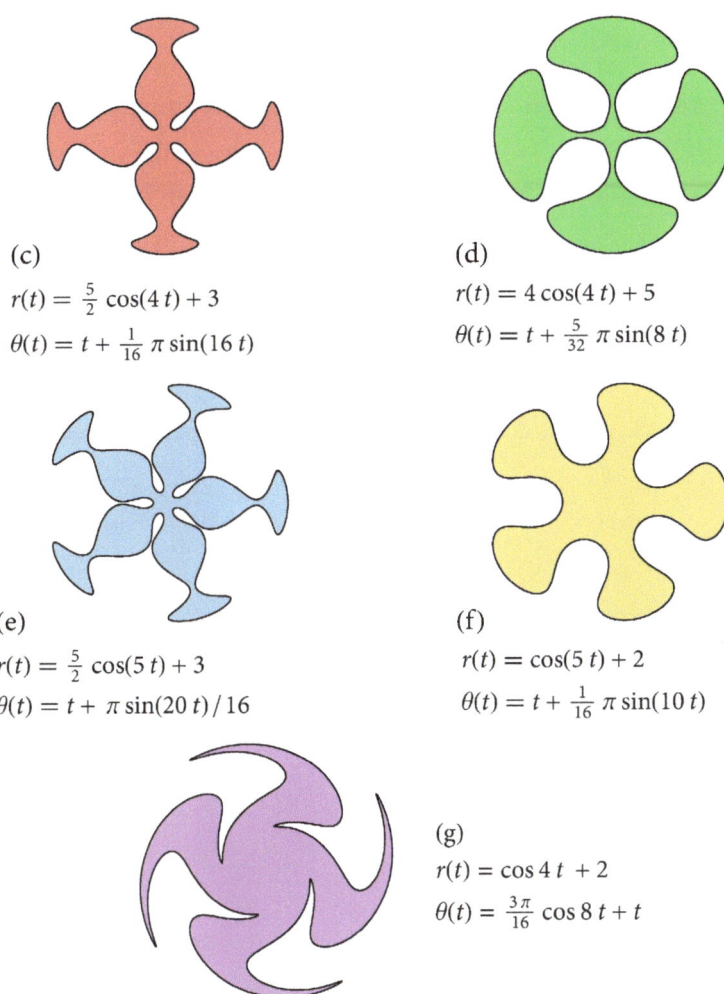

(c)
$r(t) = \frac{5}{2}\cos(4\,t) + 3$
$\theta(t) = t + \frac{1}{16}\pi\sin(16\,t)$

(d)
$r(t) = 4\cos(4\,t) + 5$
$\theta(t) = t + \frac{5}{32}\pi\sin(8\,t)$

(e)
$r(t) = \frac{5}{2}\cos(5\,t) + 3$
$\theta(t) = t + \pi\sin(20\,t)/16$

(f)
$r(t) = \cos(5\,t) + 2$
$\theta(t) = t + \frac{1}{16}\pi\sin(10\,t)$

(g)
$r(t) = \cos 4\,t + 2$
$\theta(t) = \frac{3\pi}{16}\cos 8\,t + t$

This last one (g) is a little bit different from the others because of the presence of the sharp cusps, which lack the same symmetry. Notice that θ varies with the cosine instead of the sine of t.

35

15 Polar Plots

Each of Figures 15-1 through 15-10 is a set of six curves related by a parameter k, which takes on the six values in the set $A = \{\frac{1}{16}, \frac{3}{16}, \frac{5}{16}, \frac{7}{16}, \frac{9}{16}, \frac{11}{16}\}$. In each case the set C represents the union of the six graphs of functions g_k. In particular, let $F(\theta) = (\cos\theta, \sin\theta)$, and $C = \bigcup_{k \in A} G_k$ such that $G_k = \{g_k(\theta) : \theta \in [0, 2\pi)\}$. The plots also list two functions r and h, and it might be helpful to consider r as the base curve and to consider h as a modifying function added in varying degrees with respect to k. Some figures have a function f instead of r, but its purpose is basically the same.

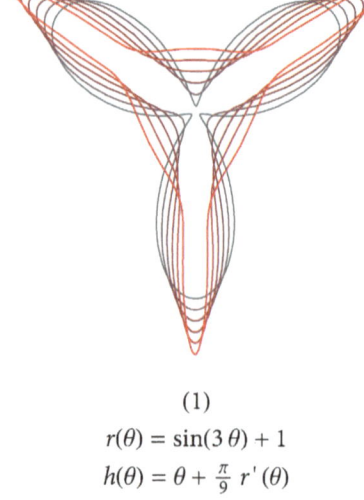

(1)

$r(\theta) = \sin(3\,\theta) + 1$

$h(\theta) = \theta + \frac{\pi}{9}\, r'(\theta)$

$g_k(\theta) = r(\theta)\, F(\theta) + k \cdot (F \circ h)\,(\theta)$

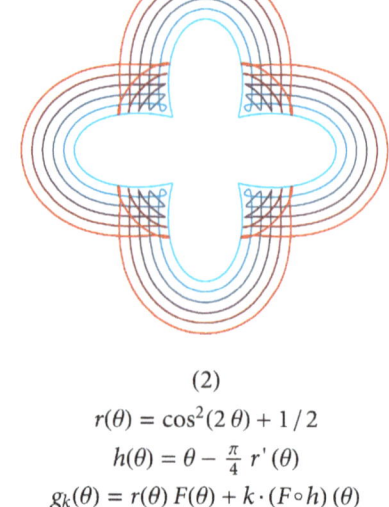

(2)

$r(\theta) = \cos^2(2\,\theta) + 1/2$

$h(\theta) = \theta - \frac{\pi}{4}\, r'(\theta)$

$g_k(\theta) = r(\theta)\, F(\theta) + k \cdot (F \circ h)\,(\theta)$

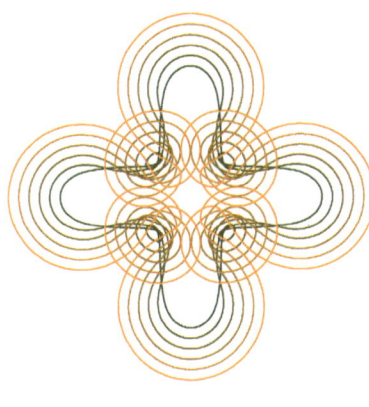

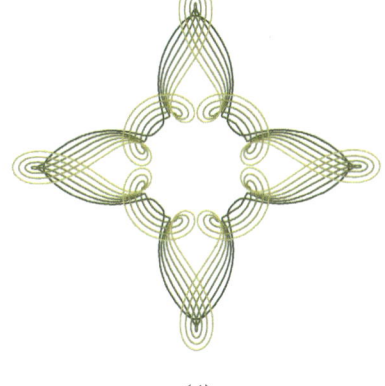

(3)

$$r(\theta) = \cos^2(2\,\theta) + 1/2$$
$$h(\theta) = \theta - \tfrac{\pi}{2}\, r'(\theta)$$
$$g_k(\theta) = r(\theta)\, F(\theta) + k \cdot (F \circ h)(\theta)$$

(4)

$$f(\theta) = F(\theta) \cdot (2 + \cos(4\,\theta))$$
$$h(\theta) = \theta - \pi \sin 4\,\theta$$
$$g_k(\theta) = f(\theta) + k \cdot (F \circ h)(\theta)$$

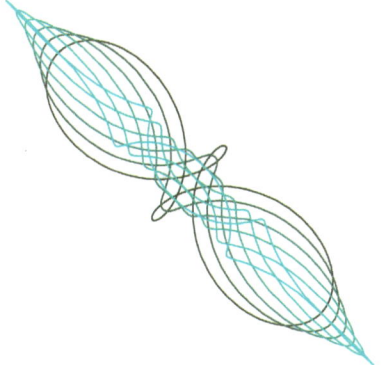

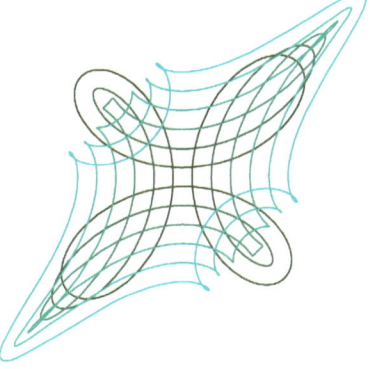

(5)

$$f(\theta) = F(\theta) \cdot (1/2 - \sin(2\,\theta))$$
$$h(\theta) = \theta - \tfrac{\pi}{2} \cos 2\,\theta$$
$$g_k(\theta) = f(\theta) + k \cdot (F \circ h)(\theta)$$

(6)

$$f(\theta) = F(\theta) \sin(2\,\theta)$$
$$h(\theta) = \theta + \tfrac{\pi}{2} \cos 2\,\theta$$
$$g_k(\theta) = f(\theta) + k \cdot (F \circ h)(\theta)$$

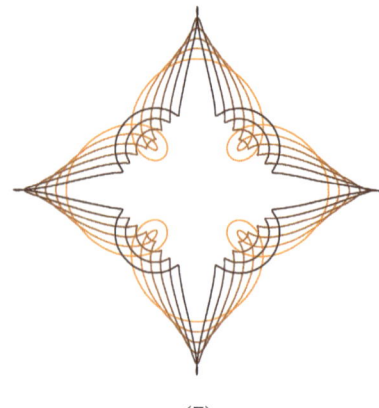

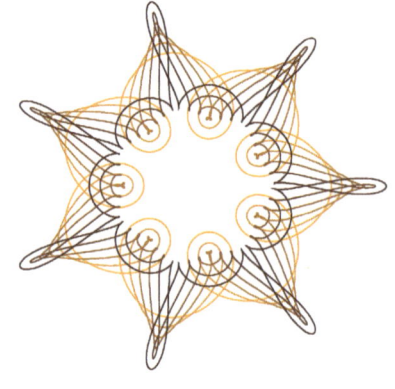

(7)

$f(\theta) = F(\theta) + F(5\,\theta)\,/\,2$

$h(\theta) = \theta + (\frac{\pi}{8}) \cdot \frac{d}{d\theta}\,(\,|\,f(\theta)\,|^2\,)$

$g_k(\theta) = f(\theta) + k \cdot (F \circ h)\,(\theta)$

(8)

$f(\theta) = F(\theta) + F(8\,\theta)\,/\,2$

$h(\theta) = \theta + (\frac{\pi}{14}) \cdot \frac{d}{d\theta}\,(\,|\,f(\theta)\,|^2\,)$

$g_k(\theta) = f(\theta) + k \cdot (F \circ h)\,(\theta)$

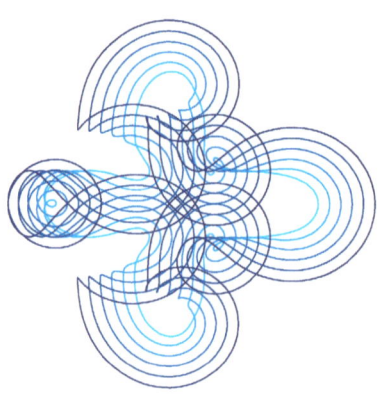

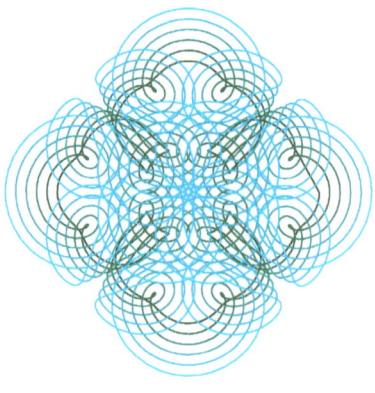

(9)

$r(\theta) = \cos^2 2\,\theta + 1\,/\,2$

$h(\theta) = -\pi\,r'\,(\theta)\,/\,2$

$g_k(\theta) = r(\theta)\,F(\theta) + k\,(F \circ h)\,(\theta)$

(10)

$f(\theta) = F(\theta) + F(5\,\theta)\,/\,2$

$h(\theta) = \theta - (\frac{\pi}{2})\,\frac{d}{d\theta}\,(\,|\,f(\theta)\,|^2\,)$

$g_k(\theta) = f(\theta) + k \cdot (F \circ h)\,(\theta)$

16 Triacontahedron

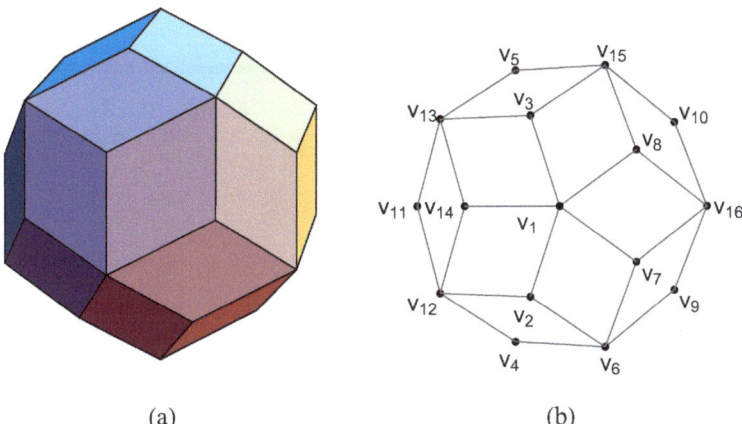

(a)   (b)

Figure 16-1

V	X	Y	Z	V	X	Y	Z
1	0	0	$c \div (2\,b)$	9	$e \div (2\,a)$	$-1/2$	$f \div (2\,a)$
2	$-f \div (2\,a)$	$-1/2$	$e \div (2\,a)$	10	$e \div (2\,a)$	$1/2$	$f \div (2\,a)$
3	$-f \div (2\,a)$	$1/2$	$e \div (2\,a)$	11	$-c \div (a\,b)$	0	$f \div (2\,a)$
4	$-b\,d \div (4\,a)$	$-g/4$	$f \div (2\,a)$	12	$-e \div (2\,a)$	$-1/2$	$c \div (2\,a\,b)$
5	$-b\,d \div (4\,a)$	$g/4$	$f \div (2\,a)$	13	$-e \div (2\,a)$	$1/2$	$c \div (2\,a\,b)$
6	$b\,d \div (4\,a)$	$-g/4$	$c \div (2\,a\,b)$	14	$-d \div (a\,b)$	0	$e \div (2\,a)$
7	$c \div (2\,a\,b)$	$h/4$	$e \div (2\,a)$	15	$d \div (2\,a\,b)$	$g/4$	$c \div (2\,a\,b)$
8	$c \div (2\,a\,b)$	$-h/4$	$e \div (2\,a)$	16	$c \div (a\,b)$	0	$c \div (2\,a\,b)$

$$a = \sqrt{5} \qquad b = \sqrt{2} \qquad c = \sqrt{5 + \sqrt{5}} \quad d = \sqrt{5 - \sqrt{5}}$$

$$e = \sqrt{5 + 2\sqrt{5}} \quad f = \sqrt{5 - 2\sqrt{5}} \quad g = 1 + \sqrt{5} \qquad h = 1 - \sqrt{5}$$

Figure 16-1(a) shows the triacontahedron, which for the sake of construction divides into three parts. They include a top and a bottom that are very similar, and a middle section which zigzags in a loop with vertical faces. The graph in Figure 16-1(b) and the coordinates in the table describe the top part, which also becomes the bottom through the application of the linear transformation T.

$$T = \begin{pmatrix} g/4 & d/(2\,b) & 0 \\ d/(2\,b) & -g/4 & 0 \\ 0 & 0 & -1 \end{pmatrix}$$

The edges of the middle faces start at the top vertices along the frontier and are lines straight down to the corresponding bottom vertices, making a total of ten vertical edges.

The peculiar looking shape in Figure 16-2 has the same thirty-sided symmetry of the triacontahedron, but it differs apparently in being merely the form of edges without faces, along with having a kind of intertwining at the vertices (although not all of them). The strips are basically just pairs of cubic splines with the flat space between them filled in.

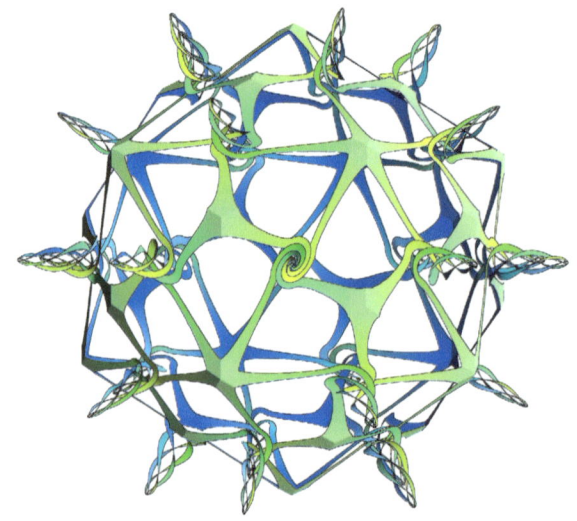

Figure 16-2

17 Cuboid

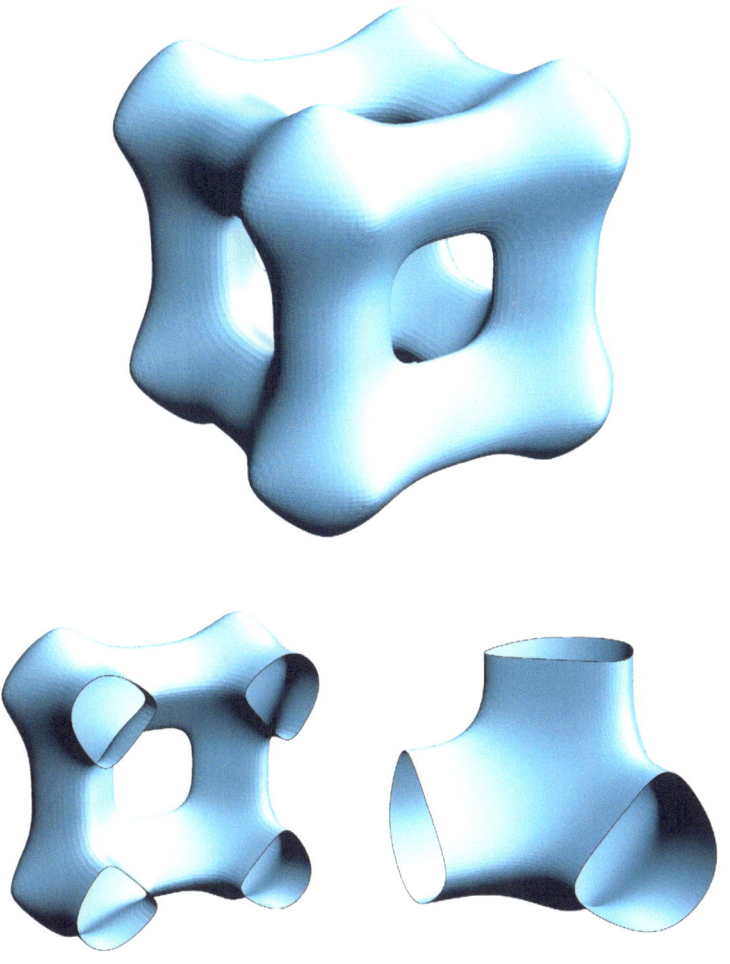

Figure 17-1

Figure 17-1 shows the solution to the transcendental equation $f = 0$ in the cube $x_i \in [-10, 10] \ni i \in 1, 2, 3$, where $f(x_1, x_2, x_3) = \sum_{i=1}^{3} h(x_i)$ and $h(t) = t^2 \cos(\pi t / 5) + 25 \cos(\pi t / 10)$. As a manifold it

seems somewhat resistant to the convenient drafting of its parameterization. On the other hand, there is a similar shape that is much easier to
handle because its components are just spheres and also a cylindrical,
hourglass shape. Let this shape in Figure 17-2 be manifold $\mathcal{M}$.

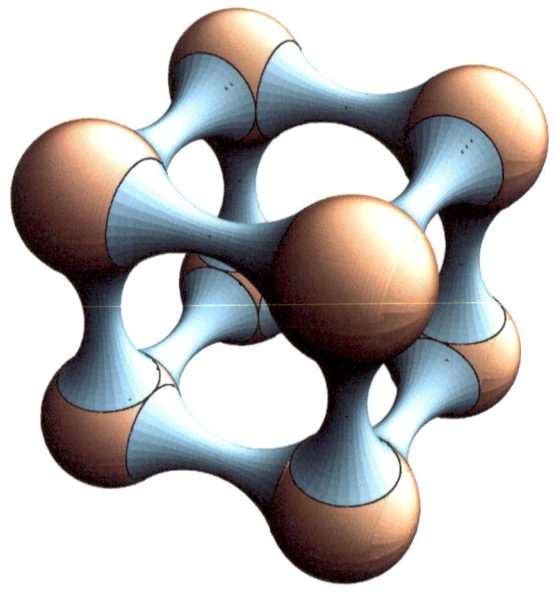

Figure 17-2

In order to parameterize $\mathcal{M}$, first note that it is a highly symmetric manifold composed of two shapes, U and V, reflected through the
various symmetries of the cube. Consider two coordinate mappings ϕ
and ψ, which may extend to the entire manifold by the actions of reflections pointwise on the mappings.

$$N_i \in \left\{ \begin{pmatrix} \alpha_1 & 0 & 0 \\ 0 & \alpha_2 & 0 \\ 0 & 0 & \alpha_3 \end{pmatrix} : \alpha_j \in \{-1, 1\}, \; j = 1, 2, 3 \right\}$$

$$U_i = N_i \cdot U, \;\; \phi_i(U_i) = \phi(N_i^{-1} \cdot U_i)$$

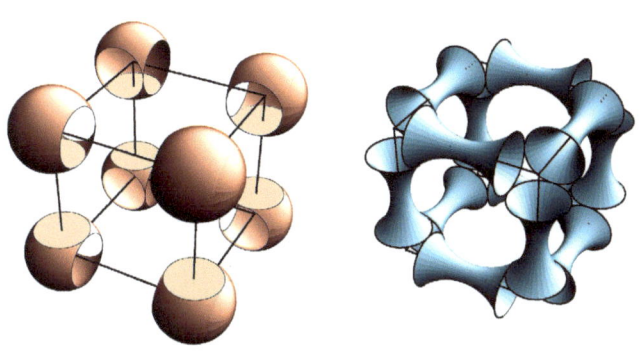

Figure 17-3

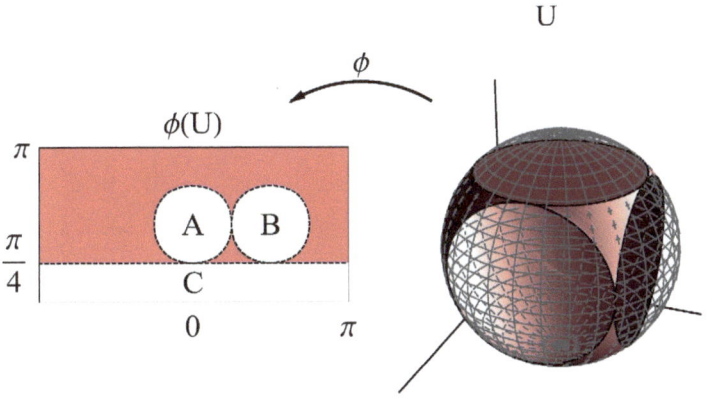

Figure 17-4

$$M_i \in \left\{ \begin{pmatrix} 0 & \alpha & 0 \\ 1 & 0 & 0 \\ 0 & 0 & \beta \end{pmatrix}, \begin{pmatrix} 0 & 0 & \alpha \\ 0 & \beta & 0 \\ 1 & 0 & 0 \end{pmatrix}, \begin{pmatrix} 1 & 0 & 0 \\ 0 & \alpha & 0 \\ 0 & 0 & \beta \end{pmatrix} \right\}_{\alpha, \beta \in \{-1, 1\}}$$

$$V_i = M_i \cdot V \quad \psi_i(V_i) = \psi\left(M_i^{-1} \cdot V_i\right)$$

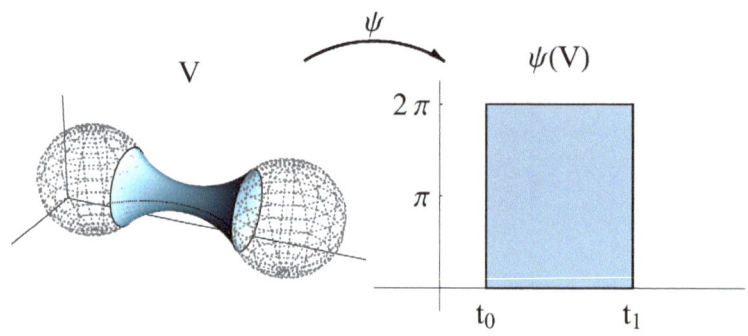

Figure 17-5

The parameterization ϕ^{-1} takes a spherical coordinate system with a translation and the removal of A, B and C. In particular, let $\phi^{-1}(x_1, x_2) = r \cdot (\cos x_1 \sin x_2, \sin x_1 \sin x_2, \cos x_2) + \boldsymbol{p}_\phi$, with

$A = \{\boldsymbol{x} \in \phi(U) \mid 1/\sqrt{2} < \cos x_1 \sin x_2\}; \quad r = -13/30 + 1/\sqrt{2}$

$B = \{\boldsymbol{x} \in \phi(U) \mid 1/\sqrt{2} < \sin x_1 \sin x_2\}; \quad \boldsymbol{p}_\phi = (\frac{-1}{2}, \frac{-1}{2}, \frac{-1}{2})$

$C = \{\boldsymbol{x} \in \phi(U) \mid 1/\sqrt{2} < \cos x_2\}$

The parameterization ψ^{-1} takes a cylindrical coordinate system with (radius) function f,

$$\psi^{-1}(x_1, x_2) = (x_1, f(x_1) \cos x_2, f(x_1) \sin x_2) + \boldsymbol{p}_\psi$$

$$f(x_1) = 1 - \sqrt{\frac{169}{900} - (x_1)^2}; \quad \boldsymbol{p}_\psi = (0, \frac{-1}{2}, \frac{-1}{2})$$

18 Modules

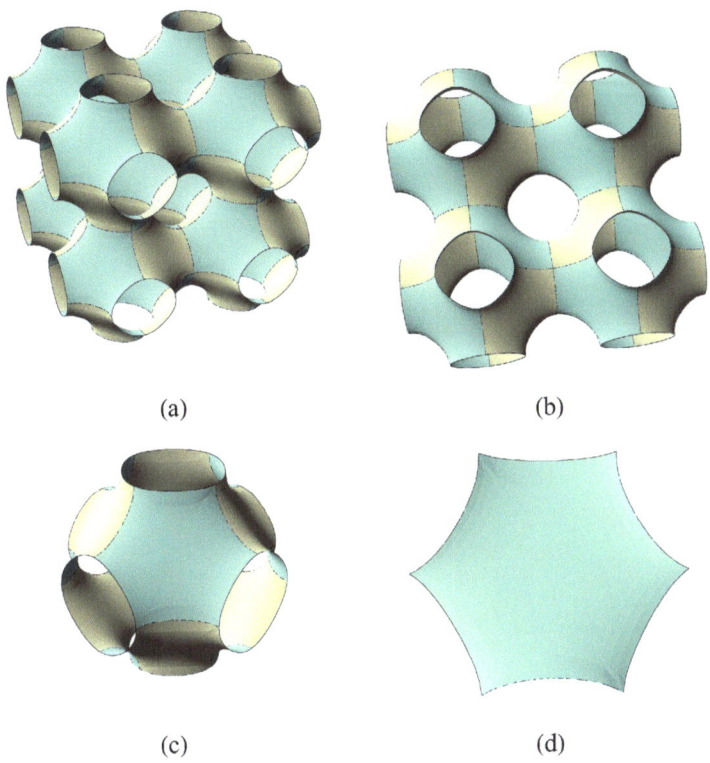

(a)   (b)

(c)   (d)

Figure 18-1

Figure 18-1 shows different views of the equation, $\cos(x) + \cos(y) + \cos(z) = 0$. The simplest piece (d) is basically like a hexagon warped in a non-planar way so that at each vertex there is a right angle. This piece then can connect to reflections of itself so to form a module that is something like a six-way pipe intersection in (c), which can in turn form larger aggregates through translation, as shown with four and eight modules in (b) and (a), respectively.

Integration and the use of symmetry help to compute the volume inside of one module. Solve the equation for z to get $z = \cos^{-1}(\cos x + \cos y)$ and integrate to find the volume under the surface; however, the region of integration must be carefully defined and takes the shape shown in Figure 18-2 as basically the shadow of the hexagon. The two curved lines in (a) are the equations $f(x) = \cos^{-1}(1 - \cos x)$ and $g(x) = \cos^{-1}(-1 - \cos x)$, and so the calculation is divided into four integrals:

$$V_1 = \int_0^{\pi/2} \int_{f(x)}^{\pi/2} z(x, y)\, dy\, dx + \int_{\pi/2}^{\pi} \int_0^{g(x)} z(x, y)\, dy\, dx +$$

$$\int_{\pi/2}^{\pi} \int_0^{\pi/2} z(x, y)\, dy\, dx + \int_0^{\pi/2} \int_{\pi/2}^{\pi} z(x, y)\, dy\, dx$$

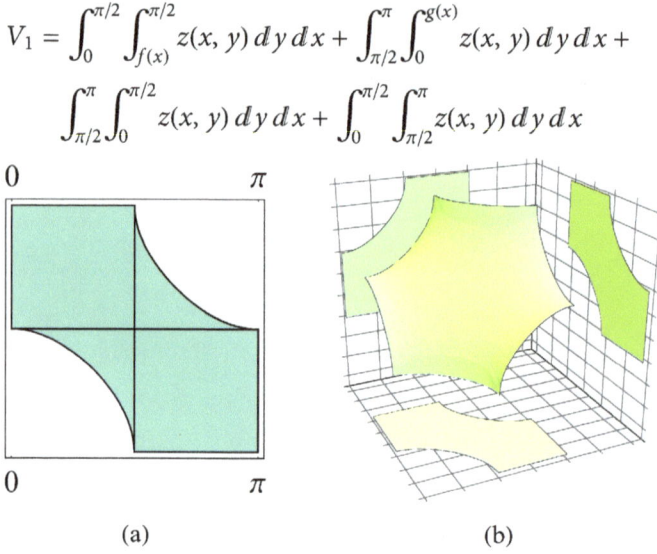

(a)   (b)

Figure 18-2

Consequential to the integration for V_1, the total volume for one module is $V = 8\,V_1 + C$ by symmetry, such that $C = \pi\, r^2\, h = \pi(\frac{\pi}{2})^2\,(2\,\pi)$, which is the volume of the central cylinder. Numerical calculation produces $V \approx 127$, which seems reasonable since for a ball with radius π, the volume is about 130.

19 Squiggle

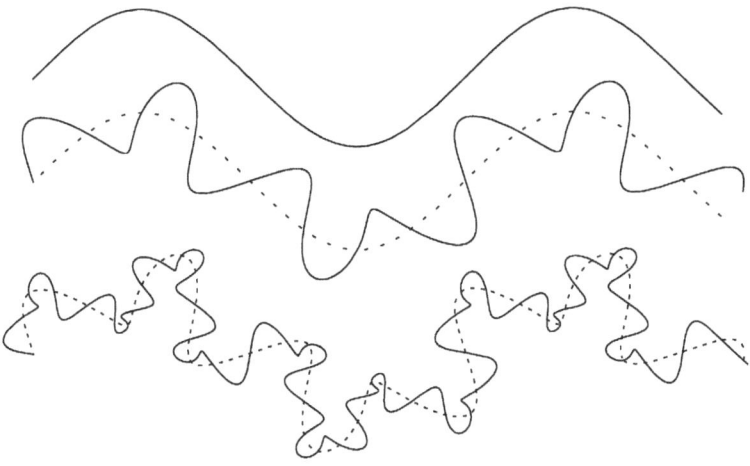

Figure 19-1

Figure 19-1 shows the three terms of a recursive sequence of periodic, parametric curves growing more and more complicated. The previous curve is drawn a second time with a dotted line. The two curves superimposed should show how one is relative to the other in the sense that the horizontal axis for graphing, although invisible, becomes a curve itself. They quickly get messy, but their definitions are nonetheless as follows. Let $a_n(t) = 0$ and $b_n(t) = 4^{-n} \sin(4^n t)$, and the x and y coordinates of each curve are a function of t.

$$\begin{pmatrix} x_0 \\ y_0 \end{pmatrix} = \begin{pmatrix} t \\ 1 \end{pmatrix} \text{ and } \begin{pmatrix} x_1 \\ y_1 \end{pmatrix} = \begin{pmatrix} \frac{dx_0}{dt} & -\frac{dy_0}{dt} \\ \frac{dy_0}{dt} & \frac{dx_0}{dt} \end{pmatrix} \begin{pmatrix} a_0 \\ b_0 \end{pmatrix} + \begin{pmatrix} x_0 \\ y_0 \end{pmatrix}$$

Then, for $n > 1$ define $\begin{pmatrix} x_n \\ y_n \end{pmatrix} \equiv \begin{pmatrix} \frac{dx_{n-1}}{dt} & -\frac{dy_{n-1}}{dt} \\ \frac{dy_{n-1}}{dt} & \frac{dx_{n-1}}{dt} \end{pmatrix} \begin{pmatrix} a_{n-1} \\ 2\,b_{n-1} \end{pmatrix} + \begin{pmatrix} x_{n-1} \\ y_{n-1} \end{pmatrix}$

The linearity of the definition has a straightforward interpretation for the form $u = T v + w$. Essentially in this case it means that u is the point found by projecting v onto basis T centered at w.

Figure 19-2 shows other squiggly curves that use nearly the same recursive sequence. The first line (a) is a graph of (x_1, y_1) and the other three, (b), (c) and (d), are graphs of (x_2, y_2) for different values of a parameter $k \in 4, 5, 6$. The functions a_n and b_n have new definitions as follows:

$$a_0(t) = \frac{3 \sin(2\, r(t)\, t)}{r(t)} \qquad b_0(t) = \frac{2 - 2 \cos(r(t)\, t)}{r(t)}$$

$$a_1(t) = \frac{3 \sin(2\, k\, t)}{2\, k} \qquad b_1(t) = \frac{2 \cos(k\, t)}{k}$$

The function r is the norm of the path derivative, $r(t) = \| (x_0{}'(t), y_0{}'(t)) \|$, and the definition of (x_2, y_2) takes a normalized matrix of derivatives.

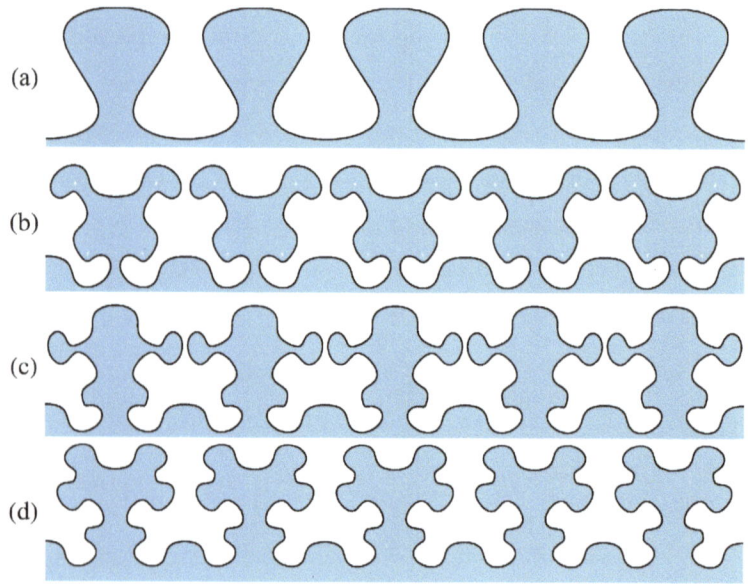

(a)

(b)

(c)

(d)

Figure 19-2

20 Concavity II

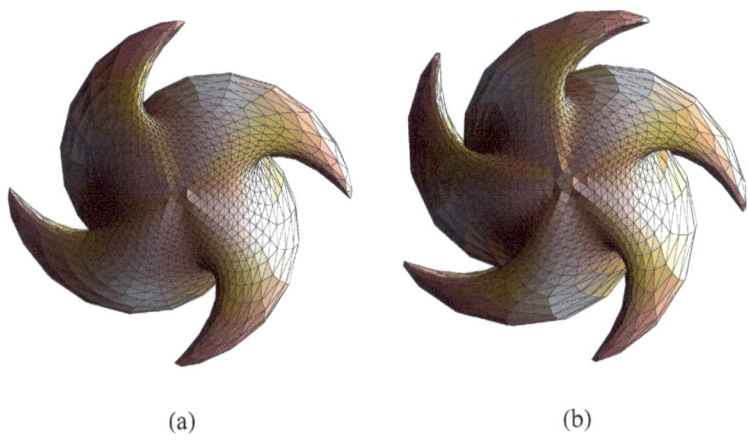

(a)   (b)

Figure 20-1

The shapes in Figure 20-1 bear some resemblance to starfish, and geometrically speaking they are part of an exercise in defining non-convex, smooth manifolds. The definitions are a challenge because a convenient coordinate system like spherical coordinates is more naturally appropriate for star-convex manifolds (such that there is a straight-line path from every point to the center). Nonetheless, with a little work it is adaptable to the present case.

The radius (distance outward) is given by r and similarly for the azimuth and elevation, by θ and ϕ, respectively. The directions make more sense if the figure is viewed from the side, shown in Figure 20-2, in which case the azimuth would be a horizontal rotation like a yaw, and the elevation would be a rotation like a change in pitch in the upward or downward direction as expected. Note that elevation is measured with zero at zenith and π at nadir.

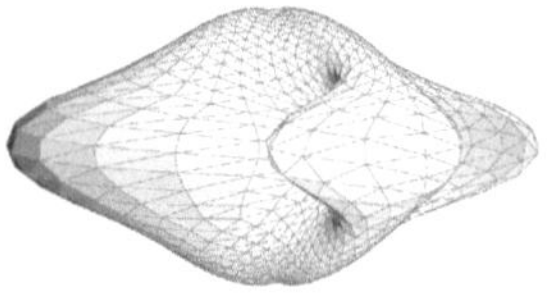

$$\begin{pmatrix} x \\ y \\ z \end{pmatrix} = \begin{pmatrix} r\cos\theta\sin\phi \\ r\sin\theta\sin\phi \\ r\cos\phi \end{pmatrix}$$

Figure 20-2

In order to define a surface in spherical coordinates, the common practice is to have the radius depend on the azimuth or vice versa, but neither would produce the cusps because of the cases in which a value of one of them must relate to at least two values of the other. Instead, the coordinates have a second set of parameters, as $r(t, s)$ and $\theta(t, s)$ with $t, s \in [0, 1]$. In particular, let $\phi = \pi s$ and let

$$\theta(t, s) = 2\pi t + \pi k(s)\cos(2\pi t g)/4$$
$$r(t, s) = k(s)\sin^2(g\pi t) + 1$$

where $\alpha = 1/100$ and $g \in \{3, 4, 5, 6\}$. The constant g controls how many cusps that the figure will have. The function $k(s)$ is a kind of bell curve with the appearance and definition given in Figure 20-3.

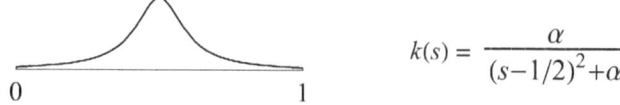

$$k(s) = \frac{\alpha}{(s-1/2)^2 + \alpha}$$

Figure 20-3

Perhaps it is interesting to ask how to compute the volume for these kind of shapes, as the integrals would require a third parameter interpreted as a depth under the surface, thus partitioning the figure into thin shells going inward. The model has however a complication in that the inner shells could not simply be smaller copies of the outer ones because then they would not fit well inside the cusps. The shells should look something like Figure 20-4 (a) and (b) with the given formulas.

$$\theta(t, s, u) = 2\pi t + (u - 1)\pi k(s)\cos(2\pi t g)/4 + u\pi$$
$$r(t, s, u) = (1 - u)((1 - u)k(s)\sin^2(g\pi t) + 1)$$

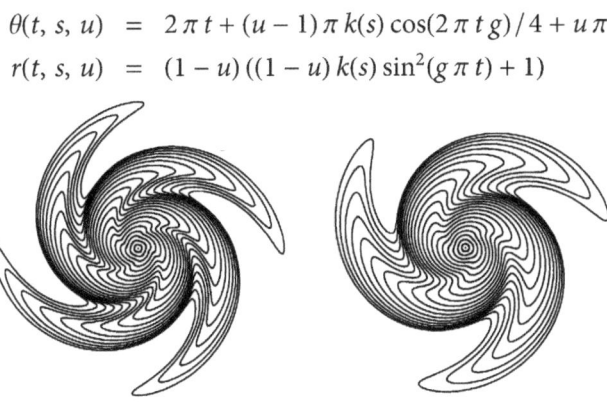

(a)   (b)

Figure 20-4

Hence, the triple integral for volume $V = \iiint dV$ is just the Jacobian over the three variables, and to attempt it any other way than by numerical computer calculation would be absurdly difficult. A series of compositions helps to define the matrix derivative;

$$(t, s, u) \xrightarrow{h} (r, \theta, \phi) \xrightarrow{f} (x, y, z)$$
$$J = D(f \circ h)(t, s, u)$$

$$J = \begin{vmatrix} \frac{\partial x}{\partial t} & \frac{\partial x}{\partial s} & \frac{\partial x}{\partial u} \\ \frac{\partial y}{\partial t} & \frac{\partial y}{\partial s} & \frac{\partial y}{\partial u} \\ \frac{\partial z}{\partial t} & \frac{\partial z}{\partial s} & \frac{\partial z}{\partial u} \end{vmatrix} \implies V = \int_0^1 \int_0^1 \int_0^1 \|J\| \, dt \, ds \, du$$

Writing down the actual expression for J is not practical since it would take several pages of equations and thus would not be of much use to see. The volume estimate with four cusps is $V = 793/100$ using numerical integration given by *Mathematica*. It seems reasonable since the estimate is larger than the unit sphere volume and smaller than that of the ellipsoid with proportion $1:2:2$.

21 Concavity III

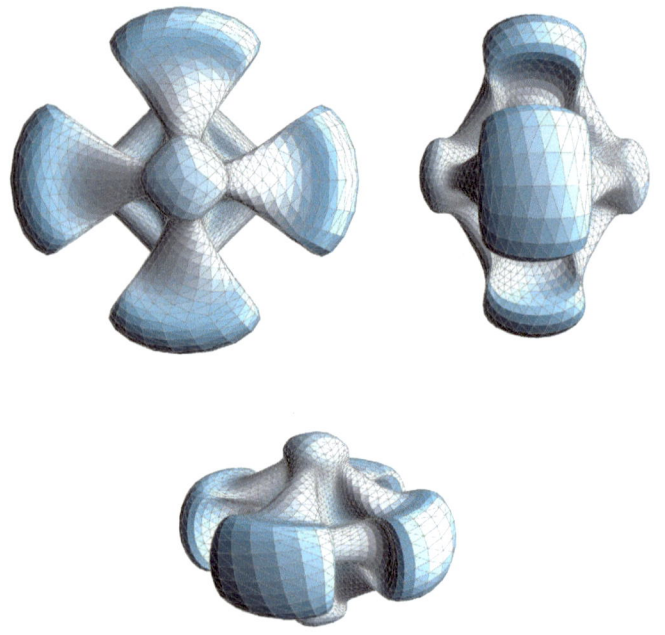

Figure 21-1

From the three views, the shape in Figure 21-1 may seem to resemble possibly a mechanical component such as a propeller or a waterwheel. Although I did not design it with any practical purpose, I did mean to make it an abstract example of a parametric surface which is continuously differentiable yet non-star-convex. So, it is similar to the shape in *Concavity II*, in that there is no point in the interior from which straight lines are extensible to all other points without the line leaving the shape. One difference between the shapes is that this one lacks convexity in two directions instead of just one.

As before, the figure takes a spherical coordinate system composed with another system specific to this space; i.e., $(t, s) \longrightarrow (r, \theta, \phi) \longrightarrow (x, y, z)$. Let $k(s) = (1 - \cos 2s)/2$ and define the two systems as shown below with $t \in [0, 2\pi)$ and $s \in [0, \pi)$.

$$\begin{pmatrix} x \\ y \\ z \end{pmatrix} = \begin{pmatrix} r \cos\theta \sin\phi \\ r \sin\theta \sin\phi \\ r \cos\phi \end{pmatrix} \qquad \begin{pmatrix} r \\ \theta \\ \phi \end{pmatrix} = \begin{pmatrix} 4 + 2\,k(s)\cos 4t + \cos 4s \\ t + \frac{3\pi}{32}\,k(s)\sin 8t \\ s + \frac{\pi}{32}\sin 8s \end{pmatrix}$$

In an attempt to clarify the meaning of the equations, consider first $r(t, s_0)$ and $\theta(t, s_0)$ with s_0 fixed. Then, the relationship between the two functions is similar to some of the pairs in *Concavity I*, with θ mostly increasing. The plots in Figure 21-2 show the level curves $(r(t), \theta(t))$ for three different values of s_0.

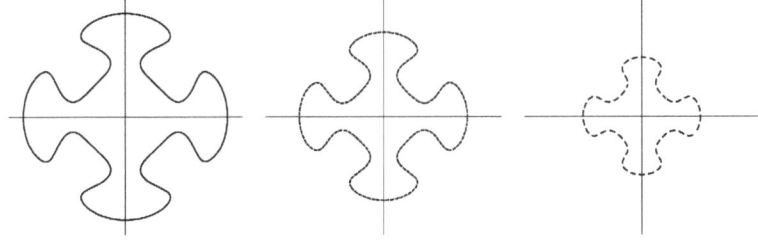

Figure 21-2

The figure grows smaller because $k(s)$ decreases while acting as the amplitude of one cosine in $r(t, s)$, and as well the concavity decreases with $k(s)$ because it diminishes the sinusoid in $\theta(t, s)$. On the other hand, looking at $r(t_0, s)$ and $\phi(t_0, s)$ with t_0 fixed reveals the changes in different vertical sections as Figure 21-3 shows, with their exhibiting clearly less concavity than in Figure 21-2.

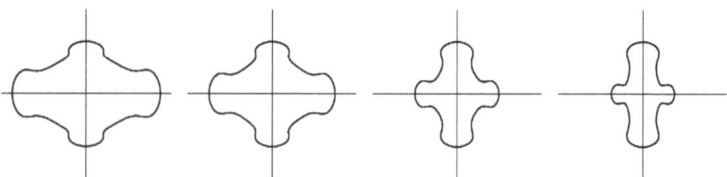

Figure 21-3

Conceivably the vertical concavity would increase with a larger factor on the sinusoid in $\phi(s)$, but kinks could begin to show in the surface with values close to $3\pi/32$ instead of $\pi/32$.

Ultimately, the shape is an accomplishment for me because it has significant concavity, but I should say that how it is is partially accidental. The four protrusions have a particular, boxed look, but when I set out to design the shape, I was expecting to have them seeming more rounded off like door knobs. Nonetheless, the boxed ones seemed more intuitive to make because θ and ϕ are perhaps more independent that way. In order to produce a more rounded look, the equation of an arbitrary circle given in coordinates of the two angles might help, but such expressions are not so obvious for any circles not parallel to the equator. Generally speaking, the spherical coordinate system preconditions calculations for meridians or for circles of latitude but not for just any circle on a sphere.

A set of other changes that are not as much trouble might include for example how the symmetry could take a clockwise bias. It could be in effect the introduction of a function $q(s) = \pi \sin(\pi s/2)/2$ meant to give the juts a bit of a smear more around the middle and less at the poles. Such a change would effect r and ϕ so that $\cos(4t)$ becomes $\cos(4t + q(s))$ and $\sin(8s)$ becomes $\sin(8s + q(s))$.

22 Self-Similarity I

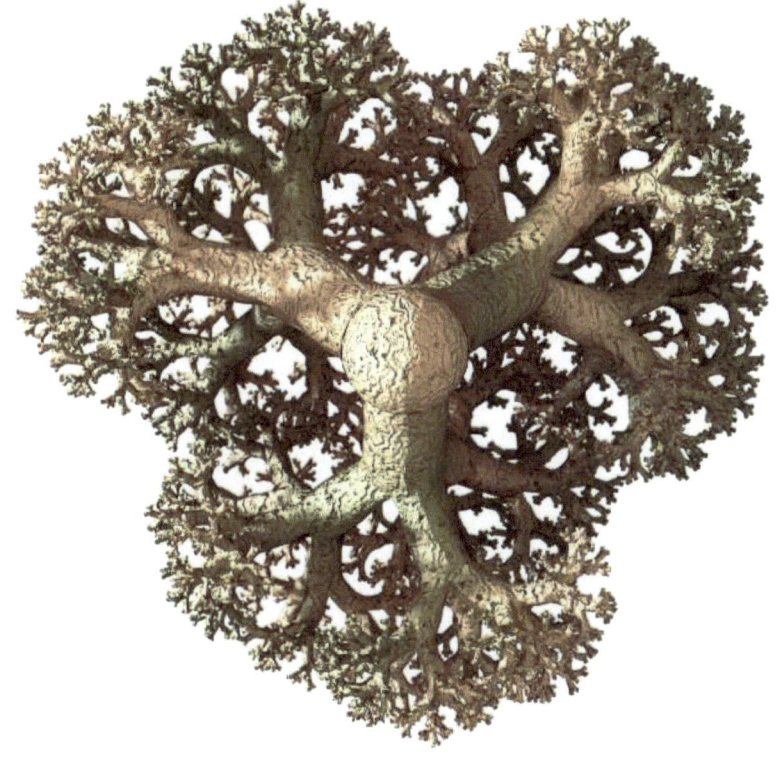

Figure 22-1

Figure 22-1 seems a bit like looking up from the trunk of a tree into its branches. Every time that the branches subdivide, they split three new ways, with the same three angles. From the circle in the middle there are three main divisions, each of which divide again into three, and so on. The division is repeatable ad infinitum, but after the fifth level, the tree starts to grow into itself. The whole of Figure 22-1 derives from a rather simple description involving many copies of the shape A shown in Figure 22-2, with linear transformations given by the three matrices.

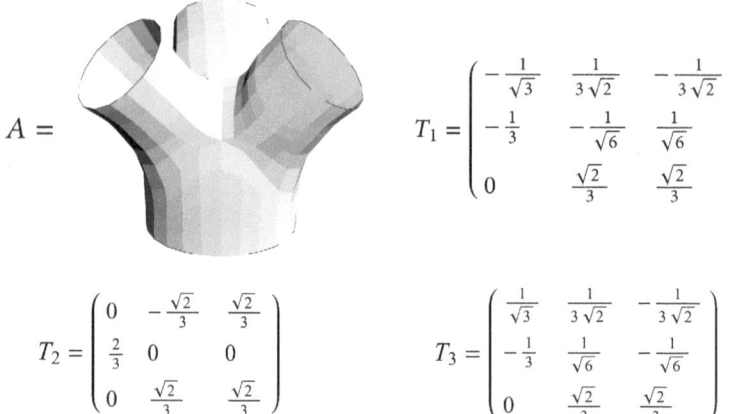

$$A = \qquad T_1 = \begin{pmatrix} -\frac{1}{\sqrt{3}} & \frac{1}{3\sqrt{2}} & -\frac{1}{3\sqrt{2}} \\ -\frac{1}{3} & -\frac{1}{\sqrt{6}} & \frac{1}{\sqrt{6}} \\ 0 & \frac{\sqrt{2}}{3} & \frac{\sqrt{2}}{3} \end{pmatrix}$$

$$T_2 = \begin{pmatrix} 0 & -\frac{\sqrt{2}}{3} & \frac{\sqrt{2}}{3} \\ \frac{2}{3} & 0 & 0 \\ 0 & \frac{\sqrt{2}}{3} & \frac{\sqrt{2}}{3} \end{pmatrix} \qquad T_3 = \begin{pmatrix} \frac{1}{\sqrt{3}} & \frac{1}{3\sqrt{2}} & -\frac{1}{3\sqrt{2}} \\ -\frac{1}{3} & \frac{1}{\sqrt{6}} & -\frac{1}{\sqrt{6}} \\ 0 & \frac{\sqrt{2}}{3} & \frac{\sqrt{2}}{3} \end{pmatrix}$$

Figure 22-2

A summary of the algorithm would be that the shape of A takes the transformation Q, which is a translation by S along with a rotation by R. The translation is $S \cdot v$ with a sum of the form $S = I + T_{i_1} T_{i_2} + T_{i_1} T_{i_2} T_{i_3} + \cdots$, such that v is just the unit vertical vector, and T_{i_k} is one of the three matrices shown above. The rotation R takes the form $R = T_{i_1} T_{i_2} T_{i_3} \cdots T_{i_n}$ for some n depending on the depth of recursion. Then $Q(x) = S v + R x$ is the image of the point $x \in A$ for a particular branch. Note that R also scales down the size of A in addition to rotating it.

More precisely, let $3_m = \{1, 2, 3\}^m$ and $S_m : 3_m \longrightarrow GL_3(\mathbb{R})$ such that $S_m(p) = (I + \sum_{n=1}^{m} \prod_{j=1}^{n} T_{p_j})$ and $p = (p_1, p_2, \ldots, p_m)$, in which I is the identity matrix. Essentially, vector p picks out a linear transformation, and similarly for $R_m : 3_m \longrightarrow GL_3(\mathbb{R}) \ni R_m(p) = \prod_{j=1}^{m} T_{p_j}$. Then, the translation S_m along with the rotation R_m are combined in $Q_m : 3_m \times \mathbb{R}^3 \longrightarrow \mathbb{R}^3$ with $Q_m(p, x) = S_m(p) \cdot e_3 + R_m(p) \cdot x$, with $e_3 = (0, 0, 1)$. The index m basically means that the image of Q_m is in the m-th level of the tree's recursion.

A tree like the one shown in Figure 22-1 would show all the branches through the sixth level (plus A itself), and so it is $C = A \cup (\bigcup_{n=1}^{6} C_m)$ such that $C_m = \{Q_m (\boldsymbol{p}, \boldsymbol{x}) : \boldsymbol{x} \in A, \boldsymbol{p} \in 3_m\}$.

The easiest way for me to compute precisely the shape of A is to use a computer-aided design tool for splines. To make a spline for the circular, bottom opening on A requires four control points. The three transformations T_1, T_2 and T_3 applied to these points determine the location of the three smaller openings at the top of A. As the twelve coordinates in all thus define the circular, boundary edges, then defining the surface is a matter of connecting the dots. The spline editor program "Hama-Patch" can export a polygonal approximation, which for linear transformations is easier to use than splines are directly. *Mathematica* calculates the aggregate figure containing $\sum_{i=0}^{6} 3^i = 1093$ copies of A, and then the aggregate goes next to the ray-tracing program *POVRAY* for colors, shading and lighting.

As a more mathematical alternative to using the design tools, which tend to be quite visual, an applicable method for finding cubic spline patches to precise specification is presented later in the section entitled *Self-Similarity II.*

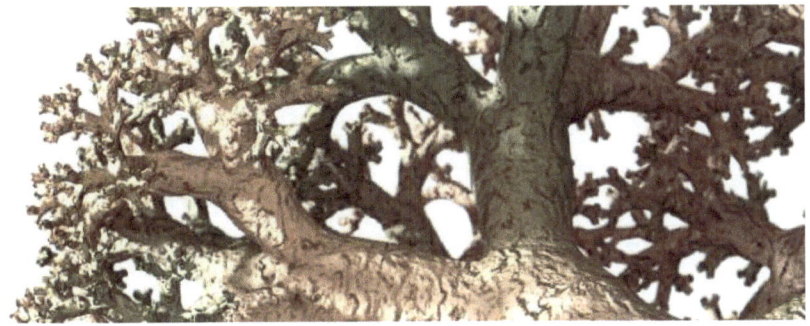

Figure 22-3

23 Labyrinthine

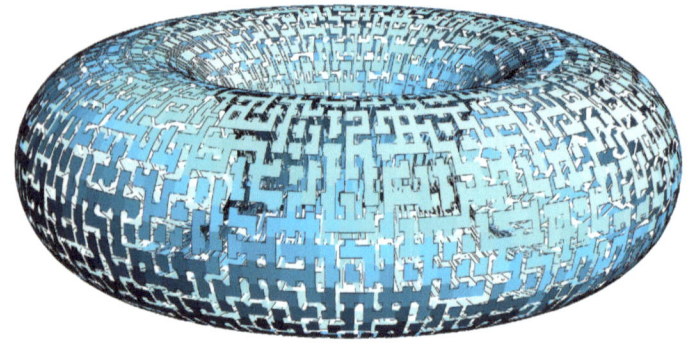

Figure 23-1

The idea with Figure 23-1 is simply to put a kind of maze pattern for a sense of transparency on a surface as an alternative to looking at solid regions. When drawing any kind of perspective figure, the computer necessarily breaks up a smooth surface into small, flat polygons (such as rectangles), and in this case the program chooses randomly from the nine puzzle pieces (in place of rectangles) in order to draw this surface. Figure 23-2 shows the basic pieces.

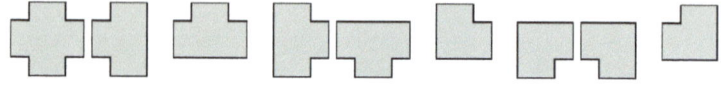

Figure 23-2

Just to be clear, the way by which the computer estimates the torus for the sake of drawing goes something like what follows. Let $F(\theta, \phi) = (x(\theta, \phi), y(\theta, \phi), z(\theta, \phi))$ be the parametric equations for a torus as given in *Torsion I*, in this case with minor radius $r = 1$ and major radius $R = 3$. For some $n \in \mathbb{N}$ define a sequence in $[0, 2\pi]$ as

$P = \{i\,\Delta\theta\}_{i=0}^{n-1}$ with $\Delta\theta = 2\pi/n$. Let $\boldsymbol{p} \in P \times P$ and for each rectangle with corner $F(\boldsymbol{p})$ define two vectors u_p and v_p as the length and width.

$$u_p = F(\boldsymbol{p} + (0, \Delta\theta)) - F(\boldsymbol{p}) \text{ and } v_p = F(\boldsymbol{p} + (\Delta\theta, 0)) - F(\boldsymbol{p})$$

Then, the rectangle at $\boldsymbol{p}$ is precisely $R_p(t, s) = \boldsymbol{p} + s \cdot v_p + t \cdot u_p$ with $t, s \in I = [0, 1]$. The drawing is the union of rectangles $\bigcup_{p \in P \times P} R_p(I \times I)$, which seems to look smoother and more like the torus as n increases.

The maze pattern goes particularly well on a torus because the surface is divisible by small rectangles. Such division is not always the case for general shapes. For the sphere in Figure 23-3(a) there tend to be very long and narrow rectangles at the poles, so it might not take the pattern very well as compared to other, more rectifiable shapes such as the rhombic triacontahedron in Figure 23-3(b). Notice how the maze pieces are many different sizes for the ball and how the rhombic image seems hardly to impart a sense of depth worthy of its real shape.

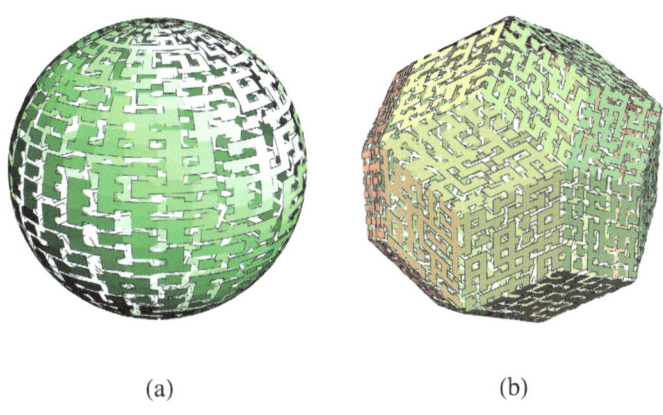

(a)   (b)

Figure 23-3

The maze idea with some adjustment might also succeed in triangulation, as it is another method that approximates smooth shapes.

24 Self-Similarity II

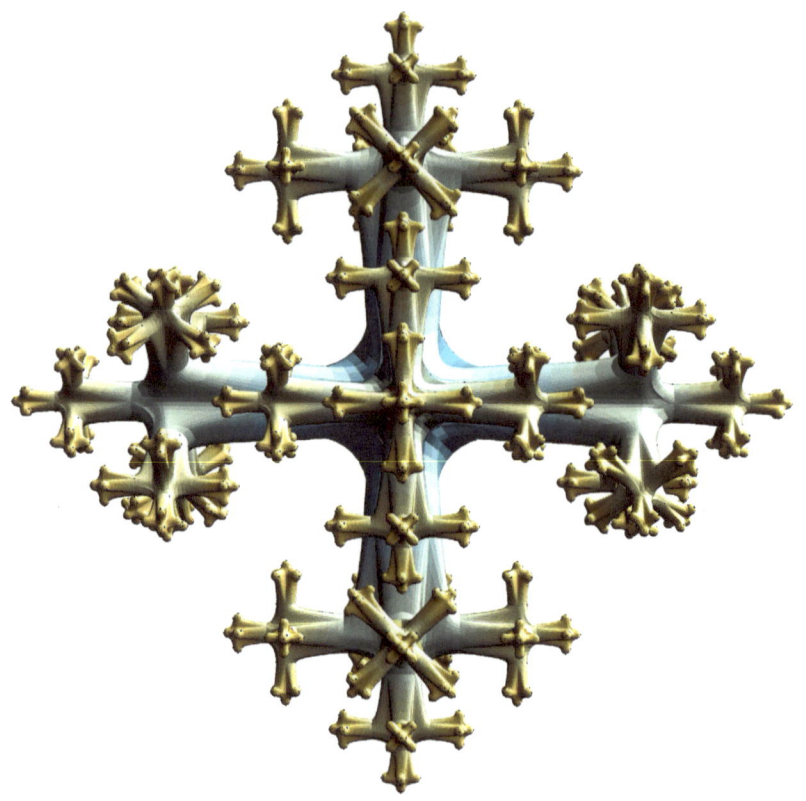

Figure 24-1

An appearance of cruciform self-similarity like toy jacks upon jacks suggests a way approaching infinity with smaller and smaller pieces in Figure 24-1. The picture is actually a view of a kind of volume which looks quite different when seen from other perspectives. The whole shape is just a set of parts, being basically a single generating object under geometric transformations of space. The object has a smooth surface defined from sets of cubic polynomials of two variables, with each set corresponding to a smaller region, and each of those has boundary splines derived from properties of their endpoints. So, in a sense the entire figure is reducible to a dozen or so points and vectors.

Let $\sigma(t)$ be a space curve for $t \in [0, 1]$ such that the coordinate functions $x(t)$, $y(t)$, and $z(t)$ are cubic polynomials of the form $a_3\, t^3 + a_2\, t^2 + a_1\, t + a_0$. It is possible to determine the coefficients by starting with the position and slope of σ at its endpoints $t = 0$ and $t = 1$.

$$a_0 = \sigma(0) \qquad a_2 = -3\,\sigma(0) + 3\,\sigma(1) - 2\,\sigma'(0) - \sigma'(1)$$
$$a_1 = \sigma'(0) \qquad a_3 = 2\,\sigma(0) - 2\,\sigma(1) + \sigma'(0) + \sigma'(1)$$

These cubic splines differ slightly from Bezier curves because the latter uses abstract control points instead of derivatives to determine the shape of the curve. Either way, a wire-frame model as shown in Figure 24-2 is a construction of several of these splines and points.

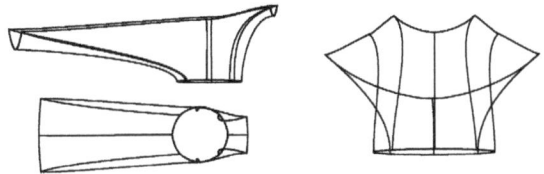

Figure 24-2

The splines and the wire-frame define a smooth, cubic surface known as a spline patch. The patch has a rectangular domain and its four edges are all spline curves. The general form of the patch is $F_k(s, t) = \sum_{i,j=0}^{3} b_{ijk}\, s^i\, t^j$, and hence there are sixteen coefficients required in the patch for each of the three spatial coordinates ($k = 1, 2, 3$). In fact, the surface is derivable from its four corners plus two slopes per corner, meaning one slope for each of the two directions (upwards and sideways, for example). Let σ_1, σ_2, σ_3 and σ_4 be the four edge splines, and let a, b, c and d represent the four corners $(0, 0)$, $(1, 0)$, $(1, 1)$, and $(0, 1)$, respectively. Note that two splines meet at each corner. The following equations form a linear system with respect to the coefficients b_{ij} and are solvable accordingly.

$F(a) = \sigma_1(0)$	$\frac{\partial}{\partial t} F(b) = \sigma_2{}'(0)$	$\frac{\partial}{\partial s} F(b) = \sigma_1{}'(1)$
$F(c) = \sigma_2(1)$	$\frac{\partial}{\partial s} F(c) = \sigma_3{}'(1)$	$\frac{\partial}{\partial t} F(a) = \sigma_4{}'(0)$
$F(b) = \sigma_1(1)$	$\frac{\partial}{\partial s} F(a) = \sigma_1{}'(0)$	$\frac{\partial}{\partial s} F(d) = \sigma_3{}'(0)$
$F(d) = \sigma_3(0)$	$\frac{\partial}{\partial t} F(c) = \sigma_2{}'(1)$	$\frac{\partial}{\partial t} F(d) = \sigma_4{}'(1)$

$\frac{\partial^2}{\partial s\,\partial t} F(a) = \sigma_1{}'(0)\,\sigma_4{}'(0)$	$\frac{\partial^2}{\partial s\,\partial t} F(c) = \sigma_3{}'(1)\,\sigma_2{}'(1)$
$\frac{\partial^2}{\partial s\,\partial t} F(d) = \sigma_3{}'(0)\,\sigma_4{}'(1)$	$\frac{\partial^2}{\partial s\,\partial t} F(b) = \sigma_1{}'(1)\,\sigma_2{}'(0)$

By sharing boundaries and corners, several patches can fit together in a way that maintains smoothness between them because the first derivatives of the surface are guaranteed to be continuous. Although, the higher order derivatives are likely not to be continuous across the boundaries. This fact is true of patches and splines alike. Geometers would say that shapes like those in Figure 24-3 have C^1 continuity everywhere and C^∞ continuity within the bounds of each patch.

Figure 24-3(a) is a kind of annulus bent up in a special way such that four copies could fit together more or less smoothly. If they are joined there is one circular hole on the bottom and five holes higher up, and the upper holes all have half the radius of the lower hole. Thus, for constructing self-similarity, five half-sized copies of the figure adjoin it in the natural way, as the larger circle becomes the smaller circles. Let the six-holed shape in Figure 24-3(b) be set A as the set of all points on the surface.

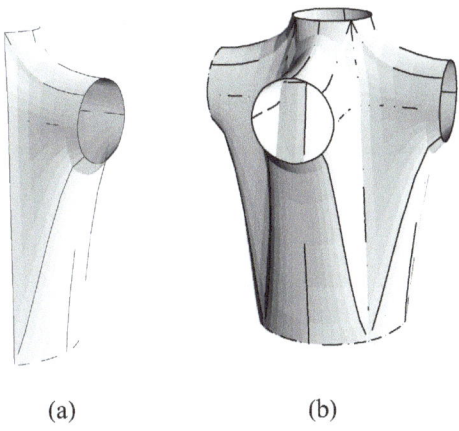

(a)   (b)

Figure 24-3

Figure 24-4 and Figure 24-5 show the self-similar construction with different levels of recursion. Although they might seem to resemble some kind of odd shrubbery, they are in fact the same surface as in Figure 24-1. Its cruciform appearance comes from a different viewpoint that is looking from directly overhead and straight down upon the shape.

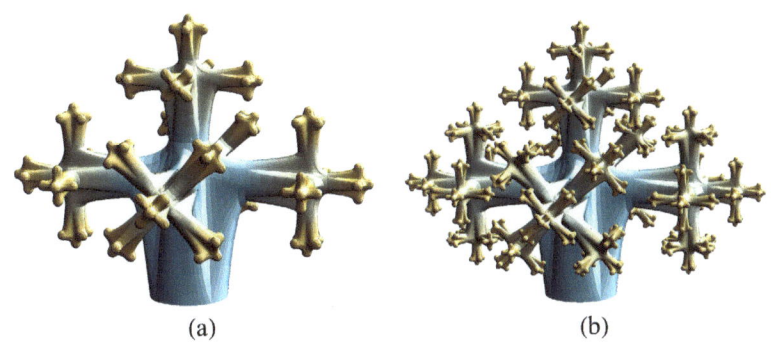

(a)   (b)

Figure 24-4

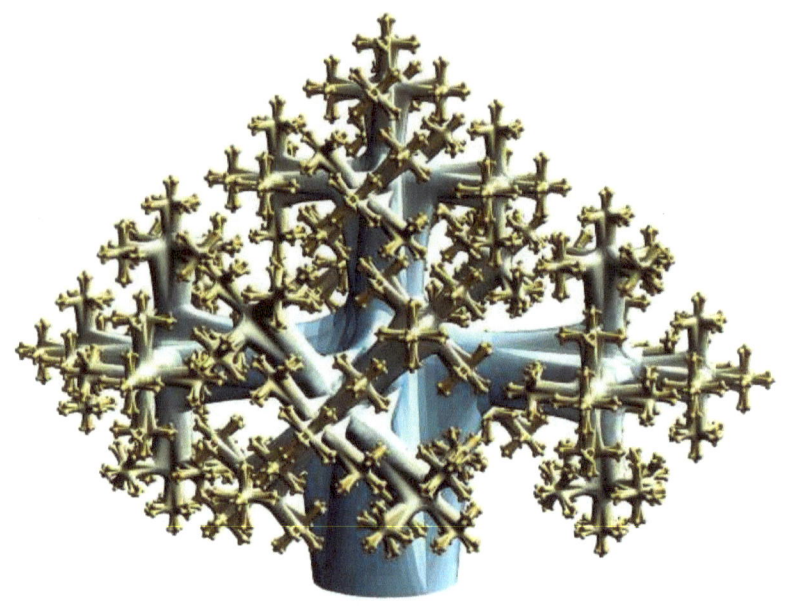

Figure 24-5

The math is a recursive application of affine transformations like in *Self-Similarity I*. For the basic figure A there are associated five spatial frames with local origins at $\{q_i\}$ and bases as matrices $\{S_i\}$ for the five small openings. For the large opening there is one more set $\{p, T\}$ including a point and a matrix. The S_i's and q_i's can be relative to p and T. The equation of the transformations would be $A' = S_i T^{-1} (x - p) + q_i$ such that $x \in A$. In this case A' is a smaller duplicate of A, and similar associations of matrices and points for A' as were made for A would lead to even smaller duplicates.

Figure 24-6 shows two overhead views with degrees of recursion less than that of Figure 24-1.

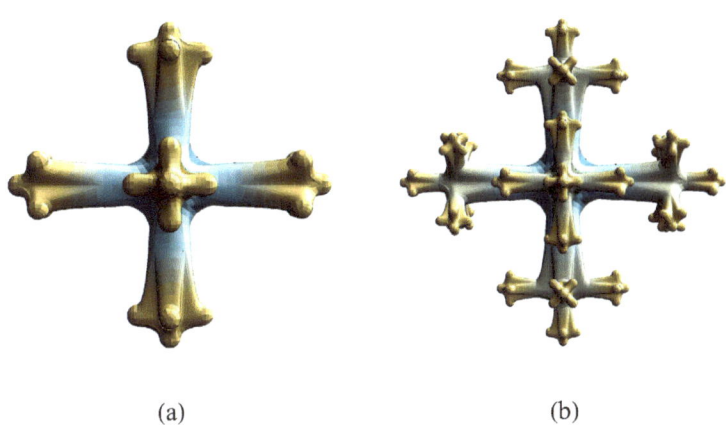

(a)   (b)

Figure 24-6

If the self-similarity of the surface continues ad infinitum such that the recursion of smaller figures is never-ending, then it may be that the total volume remains finite while the total surface area goes to infinity. Let V be the volume of A, and since with each subdivision the linear dimensions shrink to half their size, the volume decreases by a factor of eight, and because each iteration produces five copies, the volume then increases fivefold, as shown in the geometric series.

$$\Sigma V = V + \left(\tfrac{5}{8}\right) V + \left(\tfrac{5}{8}\right)^2 V + \cdots + \left(\tfrac{5}{8}\right)^n V + \cdots = \tfrac{8}{3} V$$

The surface area must reduce by a factor of four for each subdivision, but there are five new pieces, and so the ratio of the geometric series is five-fourths. Since it is greater than one, the series diverges.

References

[1] Boothy, William M. 2003. *An Introduction to Differentiable Manifolds and Riemannian Geometry, 2nd Edition.* Academic Press: San Diego.

[2] Burington, Richard S. 1973. *Handbook of Mathematical Tables and Formulas, 5th Edition.* McGraw-Hill: New York.

[3] Hungerford, Thomas W. 1974. *Algebra.* Graduate Texts in Mathematics Vol. 73. Springer: New York.

[4] Kalajdzievski, Sasho. 2008. *Math and Art: An Introduction to Visual Mathematics.* CRC Press: Boca Raton.

[5] Kepler, Johannes. (1997). *The Harmony of the World.* Translation by E. J. Aiton, A. M. Duncan, and J. V. Field, in *Memoirs of the American Philosophical Society:* Vol. 209. APS: Philadelphia.

[6] Lodö, Lama. 1982. *Bardo Teachings.* Page 69. KDK Publications: San Francisco.

[7] Lodö, Lama. 1985. *The Quintessence of the Animate and Inanimate.* Page 185. KDK Publications: San Francisco.

[8] Spivak, Michael. 1965. *Calculus on Manifolds.* Perseus Books Publishing: Cambridge.

Glossary of Symbols

○ Set Theory

$\mathbb{R}$ the set of all real numbers

$\mathbb{Z}$ the set of all integers, both positive and negative, and zero

$\mathbb{Z}^+, \mathbb{N}$ the set of all positive integers

S^2 the unit sphere; the surface of a ball with radius of one unit in length

$\mathbb{R}^3$ three-dimensional, real space

$\mathbb{C}$ the complex plane; all sums of the form $a + i\,b$ where a and b are any real numbers

$A \times B$ the product space of sets A and B; the set of pairs (a, b) with a in A and b in B

$A \cup B$ union of sets A and B

$\bigcup_{k=1}^{n} A_k$ iterated union of sets $A_1, A_2, \ldots, A_n$; e.g., $\bigcup_{k=1}^{3} A_k = A_1 \cup A_2 \cup A_3$

$\overline{A}$ means the topological closure of A

$\subset$ means "is a subset of" e.g., $\{2, 4\} \subset \{1, 2, 4\}$

$\{x : p(x)\}$ the set containing all elements which meet criteria p

$\{a, b, c\}$ the set containing elements a, b and c

$\in$ means "is an element of" a set

o **Linear Algebra**

B	bold signifies a vector
v_i	subscripts distinguish one variable from another.
$v \times w$	the cross product of vectors; a vector normal to v and w
(a, b, c)	a point in space for three real numbers a, b, and c
e_i	the unit spatial vectors $e_1 = (1, 0, 0)$, $e_2 = (0, 1, 0)$, and $e_3 = (0, 0, 1)$
$T.v, T\,v, T \cdot v$	the multiplication of a matrix T by a vector v
$GL_3(\mathbb{R})$	the set of invertible 3×3 matrices with real-valued entries
I	the unit interval $[0, 1]$ or the identity matrix
$\begin{pmatrix} a \\ b \end{pmatrix}$	a column vector if a and b are scalars; a matrix if vectors

o **Logical Symbols**

$\ni$	means "such that"
$\forall$	means "for all"
$\implies$	means "implies" e.g., $2 + x = 3 \implies x = 1$
$\approx$	means "approximately equal to"
$\oplus$	operator with variable meaning

○ **Other Notation**

x, y, z	variables often meaning length, width and height
ϕ, θ	constants or variables often meaning angles
r, R	constants or variables often meaning radii
π	the ratio of diameter to circumference of circle
$\hat{n}$	no special meaning except that $\hat{n}$ is different from n
$[a, b)$	a half-open interval
$f : A \longrightarrow B$	f is a function with domain A and range B
$f(x)$	a function named f taking one variable x, typically a real number
$f'(x)$	the first derivative of function $f(x)$
$(f \circ g)(x)$	the composition of function f with function g
$\mid A \mid$	the absolute value, norm, or the length of A
$\angle ABC$	The angle with vertex B and sides AB and BC
$1, 2, \ldots, 6$	the ellipsis means to continue the pattern; e.g., 3, 4 and 5
$3/4$	the fraction $\frac{3}{4}$; forward slash means division
(a, b)	coordinates of a point in the plane, or the open interval between numbers a and b
i	the imaginary number $\sqrt{-1}$

$\prod_{k=1}^{n}$	iterated product; e.g., $\prod_{k=1}^{3} 2k + 1 = (2(1) + 1)$ $(2(2) + 1)(2(3) + 1) = 105$
$\sum_{k=1}^{n}$	iterated sum; e.g., $\sum_{k=1}^{3} k^2 = 1^2 + 2^2 + 3^2$
$\cos(x)$	cosine function
$\sin(x)$	sine function
$\tanh(x)$	hyperbolic tangent function
$\frac{d}{dx}$	the derivative with respect to x
$\equiv$	equality in the defining sense
$\Delta\theta$	a constant meaning a small change in θ
$\| x \|$	the norm of x

About the Author

Here is a brief summary of Christopher Alan Arthur's experience involving mathematics and computer graphics. As a hobby for the high school years, he experimented with ray-tracing and geometric modeling software on his home computer and shared his animation videos with friends. At college, he studied math and worked as an engineering video illustrator, applying what he had learned with the home computer. He received a bachelors degree with honors in math and contributed to a television series about thermodynamics. A virtual reality company later employed him as a graphics programmer, focusing on animation and the interactivity of "real-time" content for amusement and business presentation. After that time he was a language teacher, a math teacher, and a student, with none of which involving graphics or programming, except that he learned to use $\LaTeX$ to write documents about math. For ten years since then, he has been drawing the pictures collected in this book. He has a website and blog related to graphics at *http://www.hythlos.org.*